Turn
Your
Ideas
into
Money

Turn Your Ideas into Money

Jeff Spira

Chilton Book Company / Radnor, Pennsylvania

Designed by Adrianne Onderdonk Dudden
Manufactured in the United States of America

Library of Congress Cataloging in Publication Data

Spira, Jeff.
    Turn your ideas into money   /   Jeff Spira.
        p.   cm.
    ISBN 0-8019-8008-9
    1. New products.   I. Title.
    HF5415.153.S67   1990
    658.5'75—dc20
                                            89-45967
                                                CIP

1 2 3 4 5 6 7 8 9 0   9 8 7 6 5 4 3 2 1 0

In memory of E.A.S.

Contents

Preface

So much has been written of late about how simple it is to take a good idea and get rich quick with it that most people think it's child's play. The truth is, few people ever profit from their ideas. It's a shame, too, because many of these ideas are excellent and, if for no other reason than that, should make money for those who put in the time and effort. Unfortunately, most of the money is made by the predators preying on those with the ideas.

In my engineering business, I often hear of and am involved with new product developers and inventors who have been fleeced by patent attorneys, model makers, supposed investors, and fast-talking "get your invention to market" salespeople. By the time I see them, they have little or no chance of ever making it. Any help I can render is too little and too late. They've either spent their entire savings, or have been convinced to proceed all the wrong way thanks to the advice of an "expert," who may or may not ever have developed a new product.

Occasionally, though, there's the individual who, through determination and persistence, makes it. One young man stopped into my office about a year ago for some advice on his new idea for a toy. He'd been recommended to me by a company I regularly consult for, so I gave my advice freely, knowing I'd get the design work from the company if he became a customer. Well, he did use the company, and I did get the design work. I hadn't heard much about it until about a week ago, when I saw the product advertised on TV! Apparently he's parlayed his business into quite a success story.

No, I haven't made a million dollars developing new products. I have, however, made a million several times over for my employers. And I do get royalties from several companies for my own products since I struck out on my own. There will be many more in the future, too.

It doesn't take a life's savings to get into the new product developing business. All it takes is the desire, a few hours of work each week, and a plan. Proceeding blindly based on advice from someone who has never actually done it, or following the traditional methods that only large corporations can afford, generally will not put money in your pocket. Most likely, just the reverse will be true.

In this book I offer a plan—a workable plan. It works for me, and it has worked for those I've coached. It can work for you, too. Putting the plan into action doesn't take any special intelligence or education—just a sincere desire and a little imagination.

Whether or not you now have an idea, this book is for you. If you don't have an idea, don't worry. After spending a little time with my exercises, you'll have all the ideas you could ever develop in a lifetime. Believe me, there is no finer feeling than walking through a store or flipping through a trade journal and seeing your brainchild proudly displayed. Somehow you get the feeling that you do make a difference.

So read on, follow your dream, and make it happen. By reading this book, you're taking the first step toward attaining your goals. May your journey be as exciting and as adventurous as mine.

1

Turn
Your
Ideas
into
Money

What ever happened to good old Yankee ingenuity? Is it gone? Where did it go? Why don't we see as much of it as we did back in the early 1900s? The answer to these questions is simple. Yankee ingenuity is here, alive and well. It still exists, yearning to be free in most people. In fact, there's more of it today than ever in the history of the world. More people than ever are keenly aware that they can make it big with a new idea.

Being an inventor used to be unique. Back in Thomas Edison's and Elias Howe's day, an inventor was something of an oddity. Most of the general public thought inventors were more than a little eccentric, yet these very same critics went right on improvising and developing products for their own use and survival, oblivious to the fact that they were actually inventing things.

Today opportunities abound for new products and new twists on existing products. An incredible number of products exists, and each product will likely have a successor. Think about the products your grandparents used when they were young. If you really think about it, you'll realize that every one of those things is obsolete now, supplanted by either a radically new product or a vastly modified product with different construction and materials. To project that concept into the future, just about every product you use today will be replaced by something new within fifteen years or so.

Well, someone has to think those things up and develop them into salable products. The manufacturing companies won't because they're too tied up putting out the products to put a lot of effort into research and development. I'm not speaking

4

here of high-tech products like supercomputers or jet fighter aircraft, but rather of the thousands of products that each of us uses every day at home, on the job, and while enjoying our favorite pastimes.

The world of new product invention and development is wide open. Everywhere you turn there's room for improvement in what we use in our everyday lives.

Do you already have an idea? Great! In this book you'll find out how to turn that idea into dollars in your pocket. If you don't already have an idea, that's OK. By the time you get through this book, you should have dozens and will know exactly what to do with them.

The biggest reason people don't take their ideas and run with them is confusion or lack of a defined plan of what it takes to turn that idea into a wallet fattener. There is a plan that works, and that's exactly why I've written this book. Most of the supposed invention plans don't work. In fact, you can spend thousands of dollars finding that out.

The classic invention development plan is to come up with the idea, locate an expensive patent attorney, then spend tens or hundreds of thousands of dollars searching for patents, generating patent texts, and paying for expensive patent draftspeople to draw up your product. You could spend years before you ever see a penny of income on the idea.

The product development plan presented here is different. It has been developed from experience and can turn an idea on a limited budget into a profit maker in a few short months. In fact, during the several months it took me to compile this book, I developed a new product, got it sold, and have already been paid for the development costs, including my time. When that product goes into production, I'll be making 10 percent of the invoice price of the product for the next ten years. The sales won't be huge because the product is very specialized, but every three months or so I'll walk to the mailbox and pull out a check for a couple of thousand dollars.

Meanwhile, on my drawing board is a commercial product that has huge sales potential. I sold it on a mere written description, and the buyer is paying for my time and expenses to develop it. I'm in for 2 percent once the product gets to market, and projected sales are in the multimillions.

Don't expect to jump right out and do this your first time, unless you already have a track record of new products in a specific area, but do expect to be able to do it once you become known and develop a few winners.

It can be done, it has been done, and with a little work and effort, you can do it, too. You can develop new products as a sideline and, eventually, as a full-time profession if you so desire.

Actually the use of the terms "inventor" and "invention" are a bit twisted by modern standards of usage. Most people consider a new product an invention only if it is patented or at least patentable. This has little to do with inventiveness and does not really affect the salability or profitability of the product.

Though my name appears as inventor on a number of patents, I make absolutely nothing on those products. The products that I do make money on are not patented and may not even be patentable. The lack of a patent does nothing to enhance or detract from their profitability or inventiveness.

A more correct term than "invention" is "new product development." New product development is actually what most "inventors" do anyway. New product development doesn't connote patents, expensive attorneys, or strange looks from your friends and family. Many new products that can end up being very profitable are actually not patentable, so they aren't really inventions. They are, however, very inventive and turn out to put lots of dollars in the developers' pockets.

Once you read Chapter 5 on the Patent Office and the true meaning of patents, you probably won't much be interested in patents either. Though I have to admit it was a thrill to see my name on a patent for the first time, I would much rather get the checks than the glory.

In my business I get a chance to see hundreds of innovative new products and new ideas every year. Very few of these items are not salable. Probably less than two in fifty are either impractical or unmakable and would never get to market. The other forty-eight can be developed and sold for a handsome profit. When you start thinking that everyone you know likely has at least one new product idea, it's surprising that so few of these ideas ever get developed. A minuscule portion of the millions of ideas ever gets past the conception stage and makes any money for the conceivers.

One of the biggest mistakes new product developers make is to think too big. They want to invent a time machine or a new method of space travel. It's fine to have those ideas, but until you have a track record in the inventing business it's far better to come up with a new idea for a toy, or a new way to restring a guitar, or an easy way to get out that tricky screw on a Chevy carburetor, or . . . or . . . . The possibilities are endless.

The procedure outlined in this book will get your product on the market. It's proven and tested. So let's roll up our sleeves and start turning that idea into money.

2

Ideas

Have

No

Value

Whatsoever

Ideas, ideas, ideas—where do they all come from? One thing is for sure, there is an endless supply of them. Many have merit, but an awful lot of them are not even worth thinking about. Americans are singularly the most new-product-idea-ridden society on planet Earth. Believe it or not, most people in other countries of the world simply do not generate a lot of ideas. New ideas are very Western, specifically American, in nature. I'm not saying that things aren't invented in other parts of the world—just that the sheer volume of ideas isn't there.

Just for kicks, I once polled an after-work drinking establishment in a smallish industrial city in Japan. The clientele was predominantly young (twenties and thirties) and mostly white-collar workers. Would you guess that no one in the entire bar had a new product idea? It was true. Now, I'm not sure if it was fair, because some simply may not have wanted to talk about their ideas, but judging from the befuddled looks and head scratching, I'd say I had thrown them for a loop.

To give credence to the test, I asked the same question in a similar establishment in a similar-sized industrial area of southern California. It had roughly the same economic and age mix as the happy hour spot in Japan. I was flooded. Virtually everyone had a new product idea. Some had three or four.

There is a reason. America has classically been the land of opportunity. Throughout our history, stories of poor immigrants making good have been told and repeated until just about all of us have been convinced that we can make a difference. We can do something new and different. We can pull ourselves up by our bootstraps. Contrast this with the prevailing idea in most other nations that you will become what your father was. Most people in other countries are taught to accept their lot in life. If their fathers were farmers or shoemakers, then they became farmers or shoemakers. People simply don't alter their class status in most countries.

Our founding fathers left us a profound and lasting gift. They insisted we become a society without class distinctions. Many people may argue with me on this point, but it's not unusual to see someone come from a poor, lower-class family and make good. I'd say that more than a third of my college graduating class at a top-notch engineering university had fathers who were blue-collar workers. In Japan, that percentage would be lower than one in a hundred. In most of Europe, it's probably one in ten.

For this reason, Americans are creative product designers. We can see something and ask ourselves how to make it better. We invent things and develop new slants on products constantly.

Ideas and How to Get Them

Many people have the mistaken idea that the only new inventions and product ideas are made in huge corporate laboratories with crowds of specialists working out the intimate details. This is flatly untrue. Every idea has a seed. One person came up with it and promoted its development.

Consider the space shuttle. This incredibly complex piece of machinery is the showpiece of American high technology. Hundreds of thousands of mathematicians, engineers, technicians, and support personnel spent probably fifteen or more years developing it, yet only one person came up with the idea. Yup, that's right. One little guy was sitting around thinking and said to himself, "Hey! how about a doohickey that can be launched into Earth orbit, fly around for a while, then glide back to Earth to make a landing so it can be used again and again." Apparently he had enough clout to convince the powers that be to develop this device, and voilà, we have a space shuttle.

For every manmade item you see in the world around you, someone thought of it first. That's pretty amazing if you think about it. Just for fun, take a look around the room or area that you're in right now. Identify all of the manmade items. Ask yourself why it was made the way it was. Why are the little notches in ashtrays? Someone had to invent that concept. Why do pencils have erasers? The answer is obvious, but it took one person to come up with that idea. A long time ago pencils didn't have erasers on one end. Why do mini-blinds have those little strings that run down both sides? Continue around the room you're in.

Take this little exercise seriously. Admittedly, many of the things you'll see are decorative rather than utile, but simply recognize a strictly decorative feature as such and continue with the useful ones. You may surprise yourself as you look around the room and give thought to the development of all the items you see. Look at these objects in depth, too. Don't merely glance at the TV remote control and decide it was developed so that couch potatoes don't have to get up to change channels. Really look at it. Check out its features. The buttons, the styling, the color, and the operation all had to be worked out. Think about the person who came up with that idea.

One thing ought to dawn on you very quickly. Every one of these items was originally developed to solve a problem. The little notches in ashtrays were developed to keep the cigarette from rolling off and burning a hole in the carpet. Remote controls for TVs were developed for invalids who couldn't get up to change the channels. Even TVs were conceived and initially developed during the Second World War so that more information could be transmitted faster. The handle on your coffee cup was invented so the inventor wouldn't burn his hands on the hot cup. Everything you see, without exception, was initially designed to solve a problem of some sort.

Toys may be the only products that are not developed to solve a problem, though that was the case when they were initially thought of. Imagine Ugh and Ooma with their new baby, Glug, in their cave relaxing after a hard day's hunting and root gathering, trying to figure out how to get their baby to quiet down. "What does the kid want now?" "I don't know—he's been fed and changed. Maybe he's just bored." "Well, we can't send him out on a saber-toothed tiger hunt, but if I paint a tiger on the wall and give him a miniature bow and arrow, he'll use his imagination."

So you see, every manmade item was conceived as a solution to a problem. Looking for ideas? Just look for problems. Every time you spot a problem, there is a potential new product idea. The problems don't have to be major and earth-shattering, either. They could be as simple as crawling under your car with the wrong size open-end wrench. Enter the adjustable (Crescent) wrench. Now you can crawl under your car with only one wrench and be reasonably sure it will fit just about every bolt you'll come across. Don't think the guy who developed that didn't make megabucks, because he surely did.

Some people insist that everything that will be invented already has been invented. In fact, the director of the Patent Office once recommended that the Patent Office be closed down. By the way, that was in 1899. Imagine all of the things that have come out since—airplanes, computers, radios, TVs, microwave ovens, even ballpoint pens! There is an infinity of more things, too, just over the horizon.

Locate a problem. Look around the room that you're in and find some small problem. Are the shoes piled up in your closet? Well, there are numerous racks and organizers for that. Spices disorganized in the kitchen? Someone invented a lazy Susan that organizes all of those little bottles. Do you have three electrical cords that need to plug into a socket with only two receptacles? Well, some astute new product developer invented plug-in socket extenders. Keep looking, though, and I'll bet you can come up with a new product idea within the next two minutes.

If you look around your house long enough, you'll find lots and lots of potential new products, but it may surprise you that many, many of these things have been developed and you just don't have them. Probably the three most popular fields for new gizmos and products are: (1) the kitchen, (2) home organizing, and (3) photography. These three areas are pretty full of enterprising inventors. I'm not saying you should avoid them, just that the competition is mighty fierce. Your hit rate with successful, sellable ideas will be mighty low.

The automotive marketplace, also, is full of innovations

and ideas. A product would have to be pretty darn special to receive any attention from an automotive accessory manufacturer. Their quantities are huge, so their tooling and manufacturing costs at risk are huge. This means that they are awfully conservative with a new product and will not even consider it unless they think the product is a sure winner.

If you've ever glanced through a photography magazine, I'm sure you were absolutely astounded at the number of gadgets available. There's room for more, too. One of my clients just invented a new type of carrying bag for all of his items. It hangs from his tripod and keeps all of his spare lenses, flashes, film, and other goodies accessible while he is on a shoot. He's having the bags made up in Taiwan, shipping them in, and doing a nice little mail-order business now.

The Workplace—The Idea Person's Dream

I often have the opportunity to walk through industrial factories. There are literally hundreds of minor problems in every manufacturing facility. I could come up with at least ten new product ideas every hour while walking through these facilities. The workplace is far, far less developed than the home.

This is how I, and many people like me, make their livings. We solve manufacturing, tooling, and efficiency problems with new product ideas. There are several advantages to this. The first is that many of these problems are unique, or at least very restrictive.

Let me give you an example. A friend of mine distributes replacement parts for machines used in photo developing labs. The machines that develop film use corrosive chemicals that play hob with many of the working parts, so they have to be rebuilt several times a year. His business is supplying the parts for rebuilding these machines.

This friend was approached several years ago by a man who invented and developed a new type of machine that checks the ingoing film to make sure the splices are all good before the film is fed into the developing machines. This amazing little product is fairly complex, uses some electronics, and sells for as much as a car. My friend agreed to distribute these machines for the foreign developer.

Well, the entire market for this machine is probably 200 machines in all of the world. There simply aren't a whole lot of large commercial film development labs that might need such an expensive device. There are probably only two or three people in the world thinking about this problem with enough ingenuity to come up with a solution. Think about the dollar potential, though. How much do 200 cars cost? Somewhere between two and three million dollars. That's plenty of profit for one or two people to make.

The workplace is pregnant with ideas for new products. Ask anyone who works anywhere what they'd need to increase their efficiency. Most people will readily volunteer a statement such as: "It sure would be easier to get my job done if we just had a thing that would _____." There's your entrée. The perfect problem for the enterprising person.

There is a down side to this, though. That is market potential. Suppose the person you asked was the only one in the world doing his or her particular job. There wouldn't be much demand for that device or invention.

Think there's no such thing? I once designed some parts for a device that would be set out on the Antarctic ice pack. Heaters would come on, and the device would melt through the ice to several miles down. As it melted the ice, the ice would freeze above it. Once it arrived at the appropriate depth under the ice, radio transmitters would begin relaying information about the characteristic of the ice. The total number to be built was three. Do you think I'd have any luck selling the ideas for that device to some manufacturer to put it into production? If you said yes, you must be a far better salesperson than I.

Most people, though, work in some capacity that has counterparts in many industries and many applications in the same industry. If someone needs a better way to unload a truck at a vitamin warehouse, chances are pretty good that there are other people at vitamin warehouses that would like that product, too. There are also people working at food warehouses, hair-care product warehouses, and all kinds of other warehouses that would find the product helpful and desirable.

Products that increase efficiency in the workplace are simple to sell. It may be difficult to sell a $500 product that will make a homemaker 10 percent more efficient, but to sell a $500 product to the supervisor in charge of a crew of ten people that will increase his efficiency 10 percent is simple. It means that he can eliminate one person from his crew. Nowadays in the United States one person doesn't have to work very long to make $500, so the supervisor's payoff is quick.

Productivity in the workplace is the key to finding lots and lots of new ideas. These products don't have to be exotic, computer-controlled, fancy machines either. I recently saw a simple little plastic device that could be filled with grease to keep ball bearings lubricated continuously. That was a masterful idea. It meant that maintenance personnel could extend the time between bearing maintenance from once a week to once every six months. In a typical medium-sized factory, you could save hundreds and hundreds of hours per year with a simple little device that sells for under ten dollars.

There is a huge vacuum of ideas in most workplaces. Watch the people in any work situation from offices to restaurants to stores to warehouses to factories to construction sites, and I'm sure the ideas will flow like water. Just watching construction workers framing a building next door to his apartment, one client of mine invented a cordless automated hammer that can drive a standard sixteenpenny nail through three two-by-fours with no effort at all. He is in the process of getting a patent and seeking a suitable manufacturer right now.

Problems, to the creative new product developer, mean opportunities. Remember this concept: Problem = Opportunity. Armed with that single equation, you can get started on a career as a new product developer.

Exercise your imagination. It takes some effort, just like playing the piano. You can't just sit in front of the piano and expect to become a concert pianist. You have to play it, play it, and play it to become good. Coming up with ideas is exactly the same. Really look at problems and practice coming up with ideas. You can do this in the shower, while driving to work, literally anytime.

Idea creation is actually structured daydreaming. Present the facts to yourself, let your mind wander, and come up with possible solutions. Don't discount any idea, no matter how silly or impractical it seems. With a little work and refinement, who knows, it may turn out to be the perfect solution to a problem.

Record every idea you have. Write it down in a notebook.

Leave yourself a complete page for each idea, and sign and date it. Look over these ideas occasionally and scrawl some new ideas, likely markets, marketing ideas, and technical ideas on the idea page that apply. You'll be amazed how quickly these add-on information pieces will fill in gaps in the idea and help refine it. When you find one that has merit and you want to develop it further, transfer it to a more formal record. I'll describe how in Chapter 5.

If you keep at it and train your mind to look at these problems, you'll have floods of ideas. You'll get ideas about how to make things better even when you're not interested in solving something.

Something else may happen as well. Your training in idea generation will likely spill over into other aspects of life. Management strategies, marketing ideas, even your love life can be affected by a creative yet logical approach to problem solving.

When you start to get good at problem inspection and idea creation, many things in your environment will begin to strike you as uproariously funny. Oftentimes I've had to contain myself from breaking out into peals of laughter looking at the way some people handle simple problems. All sorts of people from politicians on TV to gardeners start to look like actors in a comedy show. It sure injects some fun into life.

Why Ideas Have No Value Whatsoever

With all of this talk about ideas, you may wonder why I named this chapter "Ideas Have No Value Whatsoever." I did so because it's true. Ideas really do not have any value. Value for something is created out of scarcity. Gold is valuable because it's not lying around in piles in everyone's yard. If it were, it would have little, if any value.

Ideas, unlike gold, are lying around in little piles in everyone's imagination. There is an infinite supply. All you need to do is tap it. Man has an incredible capacity for conceiving ideas. I doubt very highly that you could run out even if you came up with a thousand every day.

With so many ideas around, how could they possibly have any value? The funny thing is that people think they have some value. This is ludicrous. Just because a person used a God-given function of his mind does not mean there is any value to it. After all, no one pays you for watching a sunset or making your heart beat. Why should someone pay you for an idea?

Everyone has heard someone say, "I have an idea that is worth a million dollars." Well, I have a lot of ideas, and I've never made a penny thinking them up. I could climb a mountain, sit on top for the next fifty years, come up with multitudes of ideas, and be totally broke when I climbed down.

If ideas are valueless, how does anyone make money from them? After all, this book is about how to make money from ideas. Notice that I never said ideas were worthless. Many have worth, and many don't. None, however, has any cash value. There is a secret to turning your ideas into money. Here's the secret: it's not the idea, but what you do with the idea. Ideas are valueless; products have value.

To turn your valueless idea into a valuable product, you must make it real. You have to put it into the physical world, not the world of thinking. I don't mean write it down or draw a picture of it; I mean build it. Make it work. Build a functional prototype. As long as your idea exists in your mind, it is valueless, as common as sand on the beach. When your idea exists in your hand, it is unique. There is only one of them in the whole world. Unique items have the highest value of all. Look at the Mona Lisa, the Liberty Bell, Lindbergh's airplane, and the Dead Sea scrolls. They are unique items and are all priceless.

Your idea, once real, is also priceless. It has nearly infinite value. You just have to convince someone else of its worth.

Let me give you an example: Suppose I just came up with a new idea for an automobile that would get 100 miles per gallon of gas, was as comfortable as a Cadillac, handled like a Porsche, could accelerate like a Corvette, and would sell for $5,000. It's a great idea that ought to be worth millions.

Suppose also that I called up Lee Iacocca, president of Chrysler Corporation, and told him my idea. Do you think he'd get a check for a million out in the mail the same day? No, he most certainly would not. In fact, he would most likely either hang up, break into hysterical laughter, or read me the riot act for wasting his time.

Now imagine that rather than just telling him about this car, I invited him to see such a car. What do you suppose he'd do then? He'd likely charter a jet, assemble a crew of engineers, lawyers, contract administrators, and some miscellaneous staff and be knocking down my door. I'd also bet that stashed in his briefcase would be a blank check for negotiation purposes as well.

No one ever had an idea that was worth a million dollars, but many have turned these ideas into products that were worth far more than a million dollars. That's the key. You absolutely must build your product before it has value. And before you build your product, you must engineer it.

3

Engineering Your Product

Engineers. Practically everyone knows one or at least has heard the term. Every city's Sunday paper classified section is filled with pages of ads for these people. College graduates in engineering are among the highest in demand and command the highest starting salaries of all graduates, including MBAs. What do engineers do? The vast majority of people in the United States who are not engineers have no idea whatsoever.

In truth, engineers are worth their high salaries. In fact, I know one executive vice president of a moderate sized aerospace company who insists that engineers are free to a company. It doesn't matter how much you pay them, they make you more than they cost. This is going a little overboard, but the theory is sound. Whatever salary you pay a quality engineer to go over, redesign, and analyze the appropriate manufacturing processes for your product will come back tenfold, unless you know and understand the engineering process yourself.

Actually, engineering your product yourself is not so difficult. It doesn't require years of schooling or the IQ of Einstein. All it takes is a willingness to learn, a familiarity with some reference books, and an ability to do some math. If you can handle algebra and read technical graphs, charts, and the sometimes boring technical descriptions of processes and components, chances are pretty good that you can engineer the product yourself.

In the context of new products, there are two distinct types of engineering that must be done no matter how simple the product is: design engineering and manufacturing engineering. Design engineering is simply analyzing the product for function and operation. Manufacturing engineering is determining the manufacturing processes and optimizing the design for manufacturability. If you have to do a nonprofessional job on one of these two because of time, money, or an incomplete understanding of what to do, make sure it's the manufacturing engineering. Most manufacturers interested in building a product will have their own staff, or at least a consultant, to perform this task. But it is critical that the design engineering—and preferably the manufacturing engineering as well—be done very well to have a truly salable product.

Every product needs thorough design engineering. Even the simplest of toys must be looked at for design integrity, strength, material selection, configuration, etc. If this step is not done, some major problems can crop up, and you may very well not have a product to sell at all.

Don't expect a patent attorney, your mother, or your favorite uncle who works for the phone company as a maintenance mechanic to know anything at all about engineering a product. Most likely they will lead you astray if anything. Locate an engineer who is familiar with similar types of products to look it over or do it himself.

Let me give you an example of what not to do:

A very nice man called me one day about a new product he'd invented. He wanted my company to manufacture it. According to him, it would revolutionize cleaning the tile line of pools and would sell like hotcakes. He had gone through the expense and trouble to patent the device (his first mistake, as you'll find out in Chapter 5), built and tested prototypes, and, all told, had invested something like $17,000 in developing this product. He called for an appointment saying he was flying in to southern California in several days. Since my company was interested in new products at the time, I agreed to see him.

It turned out the man had done nearly everything right as far as the logistics went. He had a pretty good idea, had built and tested prototypes, had obtained a patent, had himself armed with secrecy agreements, and made a professional-sounding appointment. There was a small problem, though. He obviously knew little of engineering.

I did sign a secrecy agreement, so I won't go into the details, but his device was powered by water running through a device to make it spin. Well, his conception of a fluid motor was based on the water wheel, an invention that was patented just a day or two after the invention of fire.

Because he elected to use one of the most inefficient devices ever conceived by man, his invention was grossly huge and heavy. It spewed water in all directions while in operation and was in general a joke.

Obviously my company wasn't interested, and we gracefully declined. Knowing the man had spent several years, thousands of dollars, and who knows how many hours dreaming of wealth on his new product, I felt sorry for him. I took him aside on the way out to his car and told him as encouragingly as I could that it needed some more work and that he ought to consider using a different type of drive motor.

Just for kicks, after he'd left, I pulled out my mechanical engineering handbook and looked up fluid motors. Yes, all of the high-efficiency fluid motors were there with diagrams, cutaway views, and formulas for performance. All this man had to do to save himself all of those thousands of dollars wasted on patent attorneys and prototype machinists was to open a book available in any library. Had he done that, he would have discovered how to make his product more efficient, lighter, and a whole lot more desirable for a potential buyer.

Had our company been interested in developing a tile cleaner, we certainly would not have picked his. We could have developed the unit easily using the proper fluid motor design and never have worried about infringing his patent.

All this aspiring product developer had done was to invest a lot of money and give every one of the companies he showed it to something to think about. A fifty dollar bill given to a mechanical engineering student at practically any college or university would have saved him a lot of time, money, and heartache.

Another man approached me recently with a new product he had developed and for which he was seeking a patent. It was an electrical product that allowed a plug-in electrical device to be locked out. It was a clever idea for limiting the amount of time your kids watch TV or making sure some appliance was not turned on when you wanted it off. Though this man was not an electrical engineer, he had researched the principles well and had come up with all of the right components and an attractive housing. His calculations, drawings, and designs were made professionally. He even sent to Underwriters Laboratories (UL) for their specification on this type of product and a list of the UL-listed components that were applicable. This, then, was a salable product.

This elusive thing called design engineering is not so hard to learn. It's merely a matter of asking the right questions. Unless you've just discovered a way to turn lead into gold, believe me, there are similar products and methods already in production and being built. Ask yourself what products are similar to this. How does the other guy approach this problem. Can I improve on his design methods?

Not everything on your product needs to be different or unique. All you're looking for is a different combination of sound, existing practices. If you have standard components and practices incorporated into your design, it will be far more salable than something that uses all custom parts and unique operating methods.

Perhaps the least understood design engineering concept is "KISS," which stands for "Keep It Simple, Stupid." The more parts you can throw away, the better. Rube Goldberg products never make it. The best engineering is the simplest. Keeping the product simple lowers manufacturing costs, meaning lower retail costs and thus a far more desirable product for the manufacturer and the customer. Simplicity usually means the product is more reliable as well. Whatever you do, keep it simple.

Back when I was a lad, the trusty old alarm clock that woke me up every morning was a masterpiece of mechanical engineering. There were hundreds of precision components connected and all working in unison. I recall taking it apart and marveling at all the little gears, levers, bearings, springs, and linkages. Recently, I had occasion to take apart a modern, digital alarm clock and was astounded at how few parts were really there. There were only two microchips. One was the clock chip, and the other the driver for the display. Aside from those two, the case was practically empty. That's keeping it simple.

Don't take this to mean that everything should be done

electronically. I recently took a consulting job to try to improve on a valve used for aircraft auxiliary power units. The valves the company was using were heavy, complex, and expensive. I managed to make it 20 percent smaller, half the weight, and one-third the cost by using the KISS method. My valve had one moving part and one spring, compared to seven moving parts, three springs, and an electric solenoid on the old design. Needless to say, the company was delighted.

Let's take a look at the actual process of design engineering used for any product so that you'll be able to sit down and do it on your own, or at least know what areas need to be addressed by the person doing the engineering.

The first step in design engineering is to make a list of the requirements. Write down what the product is to do, how much it should weigh and cost, any performance criteria, in what environment it should operate, anything that has to do with what the product needs to do or be. For example, let's take a look at a simple product such as a roof rack for an automobile.

Here's an example of what you might write down for the requirements:

1. **Must fit cars between 45 inches and 75 inches between gutters. (This will be determined by actual measurement of the skinniest and widest cars you can find. Add at least 1 inch to the widest and subtract at least 1 inch from the narrowest, just to be sure.)**

2. **Environment: outdoor—sun, rain, snow, etc.**

3. **Load capability: 500 pounds. (I picked this as an**

arbitrary number. If you wanted to carry a boat, skis, or other specific loads, you'd do this by weighing the heaviest load.)

4. Retail price, $39.95; cost to manufacture, approximately $4 (see Chapter 6).

From this list you can begin your engineering. The first step is to make a sketch of what you think this product might look like. Really think about every component. Go down to your local auto accessories store and see how the competition does every detail. Figure out what you like about them and what you don't like about them. Look for innovative areas in other products that you can incorporate into your design. Find areas you think need improvement.

This step is called configuration design. If you get stuck somewhere about how to do something, pull out the books and catalogs. Go through the *Thomas Register* (a multivolume list of manufacturers) to find makers of some of the components. Call and ask for their catalogs. Search through the trade journals for this and related products. You may also check out different trades for similar goods. For instance, you may get good ideas for ski racks from a bicycle rack manufacturer.

It may sound as if I'm saying copy, copy. No. Don't just copy. Refine, improve, and combine ideas. There's no sense reinventing the wheel. Remember, this product must have some unique feature or features—otherwise it wouldn't be a new product or invention. I'm assuming you went through the problem-solving step before you ever started engineering. It's in the details of the design that you want to be standard and conventional, not in the overall design.

From this configuration design, it's time to start looking at the individual components and how they are to be made. This requires some math, so if you're not up on your algebra, you'll probably want to find someone to do this for you.

Back to our automobile roof rack example. You must first decide what loads will affect the product. The first one to consider is the dead load, that is, how much weight it will be carrying. If it's a ski rack, it won't be much, but if it's for boats or luggage, it may be substantial. To this you'll add the dynamic load, air friction. You should calculate this out, but you may want to run a test. It doesn't take a university laboratory to run some simple tests that will prove your point. That you tested your ideas at all, no matter how crudely, will be a tremendous asset when it comes time to present your product to a potential manufacturer.

Now that you know the loads, you need to design the hardware to accommodate those loads. This is called stress analysis. There are numerous books on the topic, but basically it is a relatively simple topic unless you're designing something like an aircraft wing. First look at the attachment points and select a material that seems suitable—in the case of the auto roof rack it would almost certainly be aluminum—then calculate how thick the aluminum has to be to take the loads. Stress analysis books have pictorial examples along with the appropriate equations.

You would never design a product just to carry the expected loads and no more. A safety factor is always included when figuring the load. What safety factor to use depends on the consequences of a failure. The standard for manned aircraft is 1.5 times the maximum load. This is low because weight is such an important consideration in things that fly.

For structural products—bridges, buildings, and the like—a safety factor of 10 is used. The structure is actually designed to carry ten times the maximum load expected. For industrial products, the safety factor is generally 3 to 5.

For our roof rack a safety factor of 5 is recommended because the product would be subjected to things like vibration and some shock loads. This safety factor would make the product strong, yet probably not add too much to the cost.

There are basic reference books on just about any engineering topic which do not require a degree or even any extensive knowledge of technical subjects to understand and apply. Oftentimes these are supplied by manufacturers of products. For instance, spring manufacturers often publish design manuals for springs. Plastic resin companies publish design guides for their products. Even metal distributors publish design guides for using their products. Generally these are available free or for a nominal cost. Investigate these sources. A simple phone call will bring you a wealth of design information that could be the difference between a successful product putting cash in your pocket and an unsuccessful one putting a hole in your pocket.

For example, look at the O-ring. This is simply a black rubber doughnut used for sealing. Most of the common size and standard material O-rings cost around two cents. Practically every O-ring manufacturer, however, publishes a technical manual that runs between 30 and 100 pages. In it you'll discover how to select an O-ring, what materials are available, what materials to select for what environments, how to cut the cavities in the sealed unit to properly accept the O-ring, what O-ring standard sizes are available, and practically every other piece of information you could ever imagine for an

O-ring. These volumes took years to develop and write. They are yours for the asking. Just write a letter or call and ask for a manual.

Use these sources. Most component manufacturers will supply you with all the help you need. Just explain that you are developing a new product. That will send up a flag to the sales department. Salespeople love to get in on the ground floor when a new product is just getting started. They figure that once they get their component on the prototype and it works, most manufacturers won't bother to change brands or styles unless the costs get really out of line.

I've had salespeople bring their chief engineer to my office and give me all the free engineering help I could ever need to get their component on the prototype device. Once they gathered the information, I got a complete report with calculations, drawings, and technical information within a few days. That's their job. Lean on them for help. They'll teach you how to use their components properly on your product.

Industrial and technical product salespeople aren't the suede shoe, high-pressure types sometimes found in the retail and consumer business. They realize you'll need only one or two today, but the product will grow over years. Most are professional and technically oriented. Many are engineers, as well. They have to have a good working knowledge of their products because most of their selling is to engineers, and if they had no clue of what their products were about, most of their customers would kick them out promptly.

After you have made a preliminary selection of the basic components for your roof rack, it's time to go back and re-establish your configuration. This design is called the baseline configuration. Make changes to your baseline configuration

Put only one component part on each drawing

Blank forms available in office supply stores

All required information to make the part on the drawing

See engineering drawing or design drafting textbooks available in any library

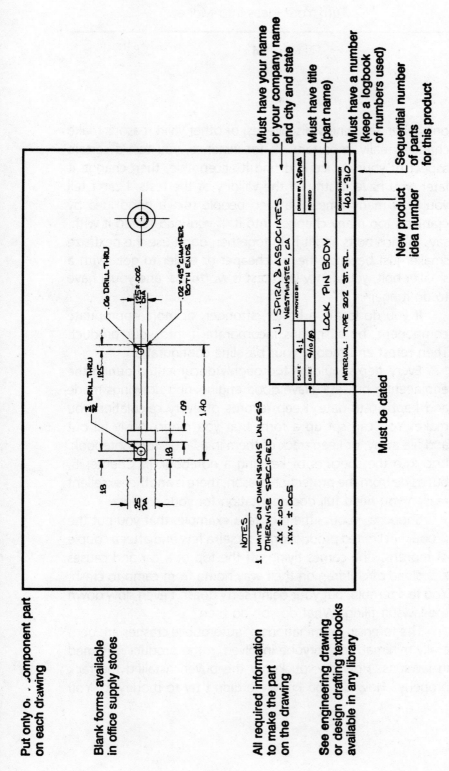

Must have your name or your company name and city and state

Must have title (part name)

Must have a number (keep a logbook of numbers used)

Sequential number of parts for this product

New product idea number

Must be dated

NOTES

1. LIMITS ON DIMENSIONS UNLESS OTHERWISE SPECIFIED
   .XX ± .010
   .XXX ± .005

J. SPIRA & ASSOCIATES
WESTMINSTER, CA

SCALE: 4:1	APPROVED BY:	DRAWN BY: J. SPIRA
DATE: 9/10/89		REVISED

LOCK PIN BODY

MATERIAL: TYPE 202 ST. STL.

DRAWING NUMBER
401-310

Sample Engineering Drawing

only when tests, new discoveries, or other valid reasons make changes really important. If, for instance, you test a certain aspect of your device and find it acceptable, then change it later, you have destroyed the validity of the tests. I can't tell you how many times I've seen people ruin a good idea by cranking too many changes into it. If you load-tested it with, say, ¼-inch bolts and it held together, don't use the next size smaller just because they're cheaper or easier to get. With a smaller bolt, your fancy load test is worthless, and you'll have to do it again.

If you do locate a better, stronger, or more appropriate component, by all means incorporate it into your product. Then retest and modify your baseline configuration.

Every step should be thoroughly documented during the engineering process. Have good engineering drawings made and kept up-to-date. Keep records of every calculation you make. You can set up a form that you meticulously fill out and file away, or keep track of them in a laboratory notebook. I go into the specifics of keeping a notebook in Chapter 5, but aside from the protection reason, there is another excellent reason you need full documentation for your product.

Suppose, in our little roof rack example, that you put the product in limited production. You sell a few and after a couple of months, one comes flying off the top of a car and causes a busload of children on their way home from camp to crash. You feel terrible, but your being sorry doesn't even slow down the lawsuit filing. What do you do then?

The forensic examinations of automobile crashes are generally minimal, so everyone involved in the product is named in lawsuits. How do you know the buyer installed the rack properly? How do you know he didn't try to modify it? You don't. If, however, you have accurate records, documented testing, and an organized system for controlling the configuration and engineering information, you'll stand a far better chance in a courtroom.

So much for the negatives. Now let's get down to some manufacturing engineering. Manufacturing engineering is simply deciding what manufacturing processes will be used to make each part. The decision is based on what process will give an acceptable appearance, strength, and function to each part.

Manufacturing engineering requires a working knowledge of the various manufacturing processes and which are most appropriate for a specific component or product. I've included many of the major processes in Appendix A along with discussions of the limitations, costs, and typical items for which the processes are used. This is intended only as a guide. Individual products have individual manufacturing needs, and any new product must be viewed independently for manufacturability.

Most, if not all, manufacturing processes are volume-sensitive. That is, a different process is used for making a few of the items than is used for high-volume production of the same item. For instance, to make the side rails for your roof rack, you may elect to have sheet metal bent into shape for the prototypes because the process is suited to small quantities. When the product goes into production, you may want to have an extrusion die made so you can extrude the finished shape. The sheet metal part will be more expensive per part, but the tooling costs are zero. In the extruded part case, the part is cheaper, but there is a fairly substantial tooling charge to make the first part.

Tooling costs for various processes must be included in

your estimates of the costs, and these are sometimes far greater than anyone imagines. This is particularly true of plastic components.

The injection molding process, used for a large percentage of plastic components, has extensive tooling costs. For example, look at the handy divider for separating the spoons, forks, and knives in your kitchen drawer. The actual cost for making one of those dividers by the injection molding process, making one part at a time, is about 20 cents. The tooling, though, costs in the neighborhood of $30,000. If you made four parts at a time, the tooling would run about $100,000, but the per piece price would be about a dime. You have to sell an awful lot of them to pay for the tooling.

What would happen if you wanted to make only a few for demonstration? You certainly wouldn't want to invest in plastic injection mold tooling. At this stage, you're not sure the product is salable or even that it will work (obviously we're not talking about the silverware divider, but something new and different). Well, you'd have to select a different process, such as vacuum forming, so that your tooling costs are only a few hundred dollars and your per piece price is several dollars.

So the crux of manufacturing engineering is first selecting the processes by which each component of your product is to be manufactured. These are just preliminary "guesses." The final selection will come as you investigate each component's manufacturability.

At this stage, you should have drawings for each of the components done. These should be real manufacturing drawings, not just sketches. Drafting is not a particularly difficult task. It can even be done on your kitchen table with a $3 T square and triangle. If you haven't had exposure to what real engineering drawings look like, a trip to your local library will yield many excellent texts on the subject.

If you feel you absolutely cannot do the drawings yourself, find a local community college with mechanical drafting classes and post a notice on the bulletin board. I assure you you'll have a flood of calls from students wanting to do your drawings for a nominal fee.

Whatever you do, do not proceed without drawings. Sketching little pictures on the back of envelopes will only get you into trouble at this stage. I've encountered many problems from a lack of drawings, and have, in fact, never seen a project succeed without them. Do the drawings.

An engineering drawing is a communication. It is in a specifically defined form to communicate every feature and aspect of the component. It tells the maker of the part exactly how the finished part is to look, what it is to be made of, and exactly what processes are required to finish the part. Without it, a shop can give you just about anything they want, and charge you for it, even if it is unusable.

A good friend of mine has a business that makes many different sizes of a similarly shaped plastic component for his products. He knew better, but told his mold maker to make a new size without drawings, figuring that because he had made several dozen of the molds in the past, the new size should pose no difficulties. He had the tool made, reserved time at the injection molding house, and had the parts shot. Well, the parts came out wrong, and he had to have the tool modified and reshot.

I was doing his engineering and drawings at the time and was charging him $25 for each drawing. He saved the $25, but then it cost him nearly $1,500 in mold modifications and molding time over what it would have cost had the tool been done to a drawing. He also lost three weeks in his manufacturing schedule for these parts. By the way, he came to me to have the drawing done later. Today, he swears he'll never again try to make anything without a manufacturing drawing.

Once you have the drawings and have at least a basic idea of what manufacturing processes will be used, you can send the drawings out for bid by local job shops. If you look through a business-to-business yellow pages for your city, available in any library, you'll find many specialty shops that produce custom parts with the processes you'll need.

The custom shops will usually offer suggestions about how to improve your design so that it can be made more easily, more cheaply, and better. It's usually better to go to the shop in person with your drawings if you're looking for advice. Oftentimes if a drawing comes to a shop in the mail, they will just bid on doing the job as is and won't help you out with improving the design. Most custom job shop proprietors have a huge amount of experience in their chosen field and can be very helpful.

Be careful, though, because the job shop is looking at only one part of your design, and their only knowledge of the product is what's on that drawing. It doesn't pay to go into lengthy discussions of how the system or invention works. Concentrate on that particular component and how it can be improved. Always keep in mind that one component is a small part of a system; you must be constantly aware of how it interacts with the system.

Your best bet is to visit a minimum of two or three job shops for each part so that you can get a variety of opinions and ideas. This helps when you look at changing your baseline configuration for manufacturability.

While you're at it, when visiting these shops, ask for a shop tour to see what's going on in their plant. The vast majority of the job shop owners are proud of their shops and jump at the opportunity to show off their facilities. While you're on the shop tour, ask a lot of questions. You'll be amazed at the wealth of valuable manufacturing engineering information that can be garnered from a well-timed, "Gee, why do you do it that way?" Just about everyone loves to talk about his specialty. Keeping your eyes and ears wide open during a tour of a job shop will net you more useful manufacturing engineering information in an hour than an entire semester in a college classroom. It will be practical, tested information too, not theory.

If you've been following this chapter, you may have realized that much of it is about letting other people help you. There is certainly no shame in this. The best engineers do it extensively. We live in such a specialized industrial world that no one person can know everything about all of the components, processes, and materials available.

The most successful new product developers use the available help extensively. They recognize that the specialty job shop foreman or the component salesperson deals daily with the problems of specifying his components or making his products. Besides, the help is free. It costs only some time, and the only effort required is asking a few questions. You certainly don't need a college degree in manufacturing engineering to do that.

It's funny that I should be spending all this time convincing you that you can avoid having your product professionally engineered because that's how I make my living. While you certainly can do it on your own, you can also botch it terribly on your own. Remember the man who invented the pool product. It cost him years and thousands of dollars for want of a few hundred dollars worth of engineering.

Engineers are professionals and can get you where you want to go a whole lot faster than you'll ever be able to research your way through. It's something akin to defending yourself in court. Sure you can spend a couple of months researching in a library to learn enough of the law and studying cases to defend yourself. The expedient way out is to hire a professional attorney, though. He can lead you past pitfalls and arrive at the desired end result far faster and with many less headaches. So it is with the engineer. He'll get your product where it needs to be far faster than if you try to go it alone.

Just as there are specialists in the doctoring and lawyering business, so there are specialists in the engineering business. There are all sorts of engineering disciplines. If you were to look up in the phone book for mechanical engineers because you had just invented a mechanical product, you'd be surprised to find out that they probably don't know anything about manufacturing processes and mechanisms. They are likely structural engineers who work on buildings, bridges, and the like.

Locating an engineer to help you out with your product may be difficult. Consultants in the product and machine design business are difficult to find. Usually they are found by word of mouth. You need to get on the phone and talk to likely users, likely job shops, and even the local university to locate a consultant or moonlighter interested in or familiar with your product.

In larger cities and industrial areas, there are businesses known as contract engineering houses. These are companies set up for the express purpose of jobbing out engineers and designers on a temporary basis. They're not cheap, but chances are they will locate a person with the exact qualifications you need. The bigger shops have a coding system for the engineers' résumés, and they search through their available manpower by computer to look for the combination of skills and experience you need.

Contract engineers are usually diligent, fast, and skilled. They're used to walking on a job and producing right away, so if your needs aren't too extensive, such an engineer may prove much cheaper than a lower-priced engineer who is less familiar with your exact needs.

Whether you elect to do the engineering yourself or have it done professionally, it must be done. It's a vital step that often separates the products that make it from the products that die an expensive and disappointing death.

4

Building Your

Product

By now you should have your idea developed on paper and be ready to proceed with the next step: building your product. This is an absolutely mandatory step. You could literally omit or mess up the rest of the techniques of developing and selling a new product and still possibly succeed at getting it sold, but any omission or big mistakes in the construction of the prototype stage will ensure failure.

People need a balance of theory and practice, ideas and matter, education and experience to really understand and be able to apply a principle. This is why colleges and universities always combine lectures and labs into required courses. A person could study physics for years without really understanding the subject until he walked into a physics laboratory and put those concepts and ideas into his real, physical world. He has to see and measure what the physics principles actually do to real-life objects before he can truly comprehend the subject.

The same principles apply to new products and inventions. You can talk about it, show pictures, mathematically prove the concept, even do computer modeling of the product, and all you'll get is a blank stare from someone you're trying to sell it to. But if you put that product into his hand, let him feel it and play with it, he'll believe it is real. He will understand the concept. He will begin to see that it is a good thing for him to get involved with.

This is why Lee Iacocca would be interested in seeing the car I described in Chapter 2 and wouldn't care a hoot about your ideas. He, like every other businessperson, needs to see proof that the idea isn't the half-baked product of a semisane mind. The real proof is a prototype of your new product idea.

There is another, very good reason you should build your product. No matter what the product and how competent the designer, often little problems crop up once you make the part. It never comes out exactly the way it was planned, and construction of the prototype is the only way you'll ever find these problems.

In the aerospace industry, this phase is called full-scale development, or FSD. You would be astounded at the number of interface and operational bugs that surface during FSD on supposedly well-engineered aircraft systems. Minor fitup problems, operational problems, and just plain mistakes in the design and engineering are caught in this phase.

Don't take this to mean that the engineering was faulty or the designer incompetent. The truth is that all engineering sciences are statistical predictions of how a system should perform and fit. Anytime you look at a statistical "average," you're opening yourself up to the problems of an item that's above average, or below average, which can really throw a wrench into the situation.

I recently developed a totally new type of fuel valve for jet engines. The valves were supposed to close at 50 psi and open back up at 45 psi. A spring was the regulating device to hold open the valve until the pressure got into the 50 psi range. I ran through all of the calculations and designed a spring that was exactly right for the job.

When we got the actual valve on the test stand, it closed at 65 psi and reopened at 60 psi. This was 30 percent off! That was unacceptable to the user, so we tore down the valve to discover why. Everything was dimensionally inspected and was within the tolerances on the design. The spring was nearly exactly correct.

After rechecking the calculations, reexamining the spring, and reanalyzing every component, we finally found the culprit. The material used to make the spring was harder than it should have been. It turned out that the material specifications call for a minimum hardness and not a maximum. It also happened that the wire manufacturer had an exceptionally good batch of spring wire so the wire was better than it needed to be to just pass the requirements, so the valve didn't perform the way it was supposed to.

These are the little details you'll need to know to make your product salable. If the buyer of the product has to discover these details by himself, you lose a lot of credibility.

Of the several hundred, or maybe thousand products I've designed from scratch, it has been a rare product for which the finished parts looked or operated exactly as I had predicted. In fact, I can recall only one design—a special machine for processing plastics—that was on line and running exactly as I predicted the day it was completed. All of the rest worked, most right away, but I was usually surprised somewhere along the way.

Again, the only way to truly understand your product is to build one. When I say build your product, I mean a working model, not a mock-up. The prototypes you build must operate functionally. It is not necessary to use all of the exact manufacturing processes and materials you plan for the production model. It's not even necessary to use the same number of parts. For the prototypes, sometimes it's easier and cheaper to make up some of the components out of several parts although they will be made out of one part in the actual production model. In any case, the components should be made of similar materials in the exact configuration and dimensions as your baseline configuration so that you can prove that the design performs.

How many prototypes should you make? It depends. If you suspect the design will require some tuning and modifications once it is built in order to perform well, one or two should be enough for you to do some functional testing. If you're sure of the design, you'll need at least two, one for "show and tell" and one to test.

Take our roof rack example. You'll want to make at least two. One you should mount on your car and start using for its intended purpose. If it is meant to carry a canoe, well, carry a canoe around on it for a while. This is functional testing. Document where, when, how long, how heavy—in short, all of the data about the product. Take some pictures, too, of the car traveling down the freeway, side streets, city traffic, dirt roads, and anywhere else you take it. This is all relevant testing information that the prospective product line buyer will be interested in.

The other prototype will be for "show and tell." You should finish this one well. Polish it up or put more attention to the finishing touches. This is the unit you'll be showing off to prospective buyers.

With some products, it's not practical or economical to build more than one. Take my automobile example that I wanted to show off to Lee Iacocca. If I were operating on a limited budget, it's likely that I couldn't afford to build more than one car. That would have to do, and it wouldn't be much of a handicap for something as big and complex as a car, yet two would always be preferable.

With many products you should build a number of them to run different sorts of tests and get varied opinions. Let's say you've developed an entirely new type of putter for golf. You might want to build ten or fifteen to give to your golf friends to solicit separate opinions on how the product works and performs.

To build your product, you should pay careful attention to the processes you select for the prototype. Most of the manufacturing processes are volume-sensitive, that is, they depend on a certain number of parts to be economical. For instance, look at a common desk-type telephone. The case for that telephone is plastic injection molded. The phone case probably costs only 50 cents in production. The tooling, however, probably costs on the order of $50,000. This means that to make one by this method would cost $50,000.50! That's a little steep for one plastic case. If you vacuum-formed the case from plastic sheet, it might cost $5, with a tooling cost of $200. This means one part will run $205. That would be about a $49,800 savings over making it exactly as it would be in production.

Appendix A lists some of the more common manufacturing processes for various materials along with typical costs of tooling and making a part. Look over these processes carefully before selecting one to make your product.

If you want to sell your product, you must build at least one to test and show off. Without that you have nothing, just valueless ideas. With a functional prototype you have a valuable product with the potential to fatten your wallet substantially. If you get nothing more from this book, at least get this.

5

Forget the Patent Office

I was quietly working away in my office a few years back, designing what I hoped would be the next generation of aircraft seals. Through the window, I could see the main part of the offices explode into activity. Salespeople who were in the office suddenly remembered important meetings, grabbed their coats, and hustled out. Secretaries either took off for an early lunch or suddenly felt the urge to go to the powder room. Draftsmen buried their heads in their books, or ran off to the blueprint room to make urgently needed copies. Though I couldn't hear what was happening, I knew from the sights that it could mean only one thing: the boss was on a rampage.

He stomped across the open area from his office to mine with the determination of a Marine Corps drill instructor ready to reprimand his troops. His face was doing its best imitation of a fire engine so I knew it was going to be a rough morning. I glanced at the clock: 11:30. I gave a silent "whew" knowing it would be only thirty minutes until lunchtime, when I could put away a few martinis if required.

"Look at this. Look at this," was all he could force out. He threw down a trade journal and stomped back to his office. I inspected our competitor's ad on the open page for some clue to what he was so fired up about. It took almost fifteen minutes of inspection before I could identify anything out of the ordinary. Aerospace hydraulic seals all look pretty much alike. There are standard designs that everyone makes and uses and some special designs that are a little different. Well on this ad there was a design I vaguely remembered, displayed as the competitor's new, super-duper design.

It turned out that our company had patented that particular design about five years before. It had been designed for a special application that was never built. We'd made and sold several dozen seals for test purposes, but it had never gone into production. I knew from the boss's tacit fury that I'd better get all of the design notes, calculations, drawings, patent applications, and patents together because he was ready for war.

I had the package of information together by 3:00 that afternoon when the team of expensive Century City (the downtown area of Beverly Hills) lawyers arrived and disappeared into the conference room.

The strategic planning and suit filing went on for six months. Even big-gun lawyers from New York were flown in for the case (Park Avenue addresses, of course). Retired designers and engineers were called back in for depositions, secretaries were put on full-time alert to shag filed information, and the entire concentration of the boss for months was consumed by the process of sticking it to the competitor who dared to design a seal like one we had patented.

When the smoke cleared, the decision was made to drop the case. The reason: we had less than a fifty-fifty chance of ever getting a nickel, and it had already cost $100,000. To pursue it further would cost an additional $200,000. These numbers don't even account for the opportunities lost by everyone putting their attention on the case and not the business at hand.

I learned an important lesson from this, and you should, too. The actual truth is that patents are far, far easier to infringe than they are to defend, and to defend them takes a lot of effort and money. The only company that comes to mind that has successfully defended their patents over the years is Polaroid, and I've heard they have more full-time law-yers on their staff than full-time salespeople.

In the case of the aircraft seal, there were three important ingredients that made the case undefendable. The first was that there was no direct evidence that the design was actually stolen. It appeared, and probably correctly, that the design was developed independently of any information about the original patented seal. If you pose the same problem to a group of engineers, chances are pretty good that at least two of the solutions will be the same, or at least very similar. In this case, the infringer had no idea that the product he'd developed was the same as the patented product.

Another ingredient of this particular case was that the patented seal was never put into production. It was proto-typed, tested, and offered for sale, but never produced and sold.

The third and perhaps most important element of this case was that the competing company was not selling this particular seal design to the same customer that the original seal was proposed to. In other words, the patent holder could not prove any financial or reputational damage because the competitor was making and selling this seal. Each sold to different companies for different applications. This is an im-portant consideration. The patent-holding company was not, in any way, damaged by the infringer.

Let's examine another patent infringement case that re-ceived a lot of press of late. A young, creative man in southern California who loved to tinker with motorcycles developed a new suspension for his off-road motorcycle. It worked great and gave him so much more control than the bike's original suspension that he decided this was the wave of the future and he'd better get his device patented. He invested his time and money to get it done, then decided to offer it to the motorcycle manufacturers.

He flew to Japan with his designs, prototypes, patents, and with all the confidence in the world that he'd soon be joining the ranks of the idle rich. All he received were polite turndowns from the manufacturers, so he returned dejected. He'd done a smart thing, though. Before revealing anything to the manufacturers, he had each person attending every meeting sign a secrecy agreement saying they would not reveal anything about the design.

About two years later, one of the manufacturers introduced their new line of dirt bikes with his exact suspension. He filed a patent infringement suit and within three years had collected $3 million dollars in damages!

This case had several things going for it. First, he could prove that he'd shown them the design because of the records of his visits. Second, he could prove that they agreed not to manufacture the design with his signed secrecy agreements, and, third, he could prove the damage done was both financial and reputational.

What was so different about this case and the first seal case? It was entirely different. In fact, in the second case, the patent wasn't even a consideration. Had the Japanese manufacturer not met with him, but rather sent away to Washington for a copy of the patent, he'd never have collected damages. If the Japanese manufacturer made some subtle changes to the design, he'd never have collected. If the Japanese manufacturer had been working on some similar design prior to his visit and could document it with dated drawings or design notes, he'd never have collected. Even if the Japanese manufacturer had subtly altered the design so that it didn't precisely comply with the claim of the inventor, he'd never have collected on the patent infringement suit.

The young man who designed and patented the new suspension wasn't necessarily doing everything right; the infringing company was just doing everything wrong. Had they just protected themselves a little, the patent infringement suit would never have gone against them. But, wait, the inventor had an ace in the hole and they knew it. They had signed secrecy agreements, then violated them. That is known as a breach of contract and, believe me, a breach of contract is far, far easier to prove than a patent infringement.

If this man had lost his patent infringement suit, which as I've mentioned, is more than a little likely, he would have grounds to file yet another suit. He could have filed a breach of contract suit, which most likely would have gone his way. The secrecy agreements, however, when used as evidence in his patent infringement suit, sealed his fate.

In the first case, the aircraft seal, the responsibility to prove wrongdoing was on the shoulders of the inventing company. They would have had to prove that the infringer did something wrong, but they could not. In the second case, the motorcycle suspension, the company had agreed to a contract, which was violated. The burden of proof was on the infringer to prove that they didn't do something wrong, which they could not do. The difference may sound subtle, but in fact there's a world of difference.

Patent infringement cases take years and years to ever get to court. Defenses are difficult, costly, and shaky. In fact, the vast majority of patent infringement cases that ever make it to court get ruled in favor of the infringer. Breach of contract suits, on the other hand, are fairly clear-cut. They are based on agreements entered into by both parties. When one party doesn't comply, well, there's your case. They tend to be bitingly fair, and any competent attorney can represent the party. You don't need a Park Avenue attorney representing you.

For this reason, a far more valuable document than a patent for a small-time entrepreneurial inventor is the secrecy agreement. The inventor, or I should say product developer, who has an ironclad secrecy agreement is far better protected than one with a patent.

Many people think patents mean protection. They mean nothing of the sort. All a patent does is offer evidence that you invented the product first. If you have a patent and someone can prove he was working on the same device before your patent was issued, the patent becomes suddenly null and void. The very concept "protected by patent" was invented by people with patents who wanted to scare away potential competitors. It has no basis in reality at all.

The real truth about patents is that they only give grounds to file suit, an expensive proposition. This is fine for big companies with attorneys on their staff and lots of money at stake already because they're willing to put up some more money to protect their stake. For the person who invents for a sideline, they do absolutely nothing. If a big corporation decides to take you on, there's very little you can do about it.

Let's take a look at what a patent really is and what it contains. I don't know how things were hundreds of years ago when perhaps patents meant something, but nowadays, patents are really a lot of fluff surrounding one paragraph. The crux of the patent and, really, all that it contains is in claim 1. The patents begin with a lot of legal mumbo jumbo, descriptions, pretty pictures, and history, but all of that is a sort of ritualized tradition. Following are the claims. Every patent has a numbered series of claims that describe the product. The only claim that has any weight at all is claim 1. Claims 2, 3, 4, and however many there are only build on and support claim 1. No new information is introduced in these claims. If someone hands you a patent, the first thing that you should examine is claim 1 because that is the only legal, infringable or defensible description of the invention or new idea. The entirety of the patent claim is in those few sentences. This is an important concept, so I'm being a little redundant and overstating it.

It's ridiculously easy to infringe legally most patents. Generally I can do it within five minutes or so. Actually only once was I stumped for more than an hour, and that was for a product that was covered by three patents, but I managed to come up with an acceptable similar product in an afternoon.

Let me give you an example from an actual patent. Claim 1, says ". . . regulating the flow by means of a tapered pin in a hole. . . ." All you'd have to do to make this exact same product is to regulate the flow by means of a straight pin in a tapered hole. It would be perfectly legal, and the product could be identical in appearance, size, and function to the patented one. Now, the inventor or assigned company could sue you, but they'd lose. Your device would be different from the patented one. It even violates claim 1 of the patent.

Does this sound unethical? Maybe it is, but it is done every day in industry. And it's perfectly legal! Products are copied every day. The inventors, by virtue of having their patents published, have taught their competition exactly how to build the products. They even give them drawings. Patents are public information. You can go to most major libraries and get a copy of every patent ever issued.

Does it sound as if patents afford any protection to the inventor? It sure doesn't to me. Having a patent is a little like standing up in a battlefield with a Day-Glo red and white target painted on your chest. The only real protection is to stay in the bushes with your camouflage paint. In other words, don't get your invention or new product idea patented. Remember, all you're doing is teaching everyone in the world exactly how to make the product.

Now that you know what not to do, how can you really protect yourself? Well, the answer to that question is not too difficult. All you need to do is not have anyone find out about your product. That sounds nice, but you'll never make any money with it if it's a secret to the whole world. You have to tell someone. The old story about mailing yourself all the original notes and sketches, then not opening the envelope is neither effective nor legal. What happens if you suddenly think of some improvements? Do you repeat the process? Which one is the legal document? It gets very muddy.

The best protection for your device, which is really just backup in case you get sued, is a three-step process. First, get yourself a notebook. This shouldn't be a loose-leaf affair, but rather a bound notebook. They can be picked up in most office supply stores. The most common are called laboratory notebooks. You'll need one for every product you're working on. Leave the first couple of pages empty for a table of contents. In the book write your notes, any calculations, sketches, etc. Starting from the back, record a chronology of every event that goes on with the product. This section should include every time you discussed the product with anyone, every time you gathered some new information about the product, every time you wrote a letter, when you performed any tests, etc.

In the body should be all design notes, drawings, test data, or any other pertinent information about the product. Sign and date every page in ink when you fill one out. Every page should also be numbered. As you make notes and sketches, start a table of contents on the front few pages that you left empty.

This book will become a record of what is known as prior art. Suppose you get an idea and work on it over the course of five years. You eventually get it worked out and sell the product to a manufacturing company. They go into production and suddenly discover that someone else patented the device last year. You and the manufacturer are sued. This notebook is a legal document and, if presented as evidence, will invalidate the patent on the basis of prior art. It is vitally important to keep up this notebook for that reason. It may not help you if you're trying to stop someone that stole your idea, but it will cover you if someone comes after you.

The next step in affording yourself some measure of protection against would-be suers is to file a disclosure document with the United States Patent Office. A disclosure document is just what its name implies: a disclosure to the Patent Office that you have invented something and intend to pursue a patent.

A disclosure document is a write-up describing your invention with a drawing or sketch. There is no formal format, but the more information, the better. The format I've developed over the years, along with instructions on how to file the document, is presented in Appendix B.

The disclosure document is good for two years. The patent

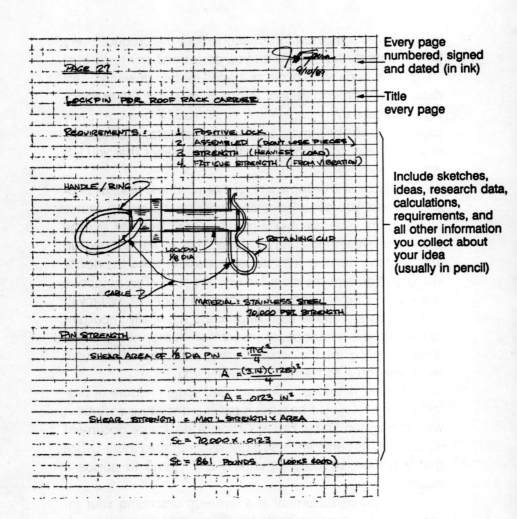

Every page
numbered, signed
and dated (in ink)

Title
every page

Include sketches,
ideas, research data,
calculations,
requirements, and
all other information
you collect about
your idea
(usually in pencil)

Sample Idea Notebook Page

office will keep it on file for that period, then destroy it unless there is further communication, or a patent application made in that time that refers to the document. Once application for patent has been made, the disclosure document gets transferred to the patent application file.

A disclosure document is not a patent, nor is it a patent application. It does not give you the right to say "patent applied for." It does, however, give you a pretty good line to use when discussing the device. "I've filed with the patent office and begun the process of getting a patent."

While issued patents are an invitation to the world to copy your device, unissued patents are a threat. No one knows what will end up in claim 1 of your patent. It's a mystery, and most people won't take the chance on possible infringement if they're not sure. Once the patent is issued, everyone will know, but an unissued patent has far, far more clout than an issued one. In fact, some patent writers are so good, they can keep the patent-pending stage going on for years and years. It pays to drag your feet in this stage.

You should wait until your product is fairly well developed and you're ready to start selling it before you file your disclosure document. It is good for only two years. You must diligently work to complete your invention and file for a patent in that time. Another little known fact is that if you offer your invention for public sale more than one year before you apply for a patent, the invention is not eligible for a patent, so hold off on selling until you have the idea well worked out and you've tested the prototypes.

In Appendix B I go into more detail on how to prepare and file a disclosure document. When you're ready, look it over carefully and go ahead, file. You'll have your name on your very own disclosure document, complete with official U.S. Patent Office number.

This chapter is titled Forget the Patent Office, so why am I telling you to go file a document with them? Well, it's not for your protection. The Patent Office will not lift a finger to protect you. Actually, the reason you want to file is solely a sales tool. With a filed disclosure document, your idea has far more value than if it is merely a developed idea or product. With it there is some scare tactic, and some intrinsic worth. Remember, much of the world still thinks patents mean protection.

The third component of protection, which is the only real protection you have, is having secrecy agreements. A secrecy agreement is a legal contract between you, the inventor, and anyone to whom you show the idea, invention, product, or device. The agreement basically states that the person who sees the product agrees not to tell anyone else about what he's seen and agrees not to build or sell the product without written consent from you.

You should get a signed secrecy agreement whether or not you have applied for a patent or filed a disclosure document. With it, you have grounds for suit even if it turns out your idea is not patentable, or for some other reason a patent is not filed for. The case is cut and dried. Imagine this courtroom scene: "You see, your honor, this company didn't make this product before. They signed an agreement that they wouldn't make it without consent from me. I showed them how, then they made it without my consent. Here are the signed agreements." What defense could they possibly come up with? The only one I can imagine is some forged papers with earlier dates. I can't imagine companies being that crooked.

You have your backup, too. All of this is entered into your lab notebook, remember?

The secrecy agreement is your friend. It's your only friend. Take it seriously. Make everyone who sees the product sign and date one, then meticulously file it where you're sure it's safe and sound. I would much rather be armed with a secrecy agreement than with a patent in a courtroom.

What can you do if someone steals your idea? If you have a secrecy agreement with him, he stole it. Immediately file a suit with a competent attorney. Also attempt to get a restraining order placed on the company to stop making and selling the product. As you'll find out in the next chapter, manufacturing a product is expensive—very expensive—so they will really squirm if you can get a restraining order put on them, and they'll be eager to settle with you. All you really want or need is a small royalty.

I've included the secrecy agreement that I've been using for years in Appendix C. It's been gone over by my attorney so I feel comfortable that it is complete and legal. Feel free to copy it, word for word, and use it yourself. You may want to have a local attorney go over it because laws do vary from state to state, and what's correct for California may not be exactly right for the state you live in.

If someone you never heard of steals your product and you can't find any connection to anyone you have shown your product to, drop it. Patent or not, just let it drop. There's plenty of opportunity out there. Go find a new idea. Chances are pretty good that the person didn't steal it from you, but rather that the idea was developed concurrently with yours. Unless you have nothing to do and plenty of money to invest over the next five years, don't bother. All it will do is cost you time and money. It may even give you ulcers. Fair? It doesn't matter. That's simply how the system works. Your efforts will be much better spent finding a new idea or product. Next time just develop it faster, sell it more aggressively, and make sure you get there first. That's how to really make it in the inventing business.

6

Why You Shouldn't Even Consider Manufacturing Your Product Yourself

Now that you have your idea that will make you millions and millions of dollars, your first plan will probably be to hang out your shingle and go into the manufacturing business. This will be the biggest mistake you'll ever make. Manufacturing takes capital, and lots of it, so unless you happen to have a couple of extra million kicking around your savings account, forget it.

It sounds incredible, but it's true. I'm not talking about setting up a huge factory and making your product from scratch, I mean any kind of manufacturing. Manufacturing takes big cash. Drill that into your memory and consciousness. I've told that to a lot of people who proceeded to look at me incredulously. "No no no," they say. "I'm planning to have someone else build it. I can have them make it for $1 and sell it for $2. I can make a clean 50 percent, so all I have to do is sell 2 million of them and then retire with a million bucks." Dream on. . . .

The truth is that even if someone else builds the product, you'll need piles and piles of cash to get your business going and keep it running. In fact, unless you have at least as much cash as your expected annual sales, you will not make it in manufacturing. That means if you plan on selling a million dollars a year worth of product, you need at least a million dollars in cash to capitalize your business, and it's still not a guarantee that you'll make it.

On this million-dollar capitalization you will be very lucky, or at least very skillful, to make 20 percent of pretax profit. After you give Uncle Sam his cut (46 percent for corporations), you get to keep about $100,000. Not a very good return on a million bucks. You'd be better off buying real estate. Not only that, but that hundred grand won't be in cash. It will be tied up in accounts receivable and inventory.

Your only savior is that you'll be building up a company that might be worth something someday—maybe. You would be astounded at the number of manufacturing companies you could buy for nothing. That's right: nothing. Just walk in and assume all of the assets and liabilities. The owners would kiss your feet just to get out from under them. In fact, very few manufacturing companies are worth the amount of money that was sunk into them. The ones that are often go public, and the ones that don't go public are rarely for sale.

Let's take a look at what it takes just to tool up a new product and introduce it into the market. Let's say you've invented a new type of can opener. It's made mostly of plastic with a metal part to pierce the cans. It is without a doubt the fastest and easiest can opener ever to come out—truly a magnificent product. Everyone you show the model to agrees, it will sell like hotcakes. You decide to set up your own man-ufacturing company to make can openers.

The first thing you need to do is tool up the product. To make it economically, let's say you have to make a four-cavity injection mold tool to get the handles for 15 cents each. Some typical tooling costs are shown below:

Injection mold tool (four-cavity)	**$25,000**
Stamping tooling for steel part	**19,500**
Automated assembly machine	**36,000**
Fixtures for assembly machine	**8,500**
	$89,000

That's not so bad. I've been involved with products that were triple that cost for little manufacturers with eight employees.

The next step is to get an office, phone, some business licenses, resale permits, corporation papers, office furniture, and the like:

Attorney's fees	**$2,500**
Accountant's setup fees	**1,200**
Office and shop deposit	**2,000**
Phone and utilities setup	**1,500**
Licenses and fees	**800**
Office furniture	**1,800**
Copier, fax, typewriter, etc.	**2,400**
	$12,200

This isn't for anything fancy, just a little 1,200-square-foot shop and office where four or five people can work. You may look at that and say I'm all wet, but try giving your local Ma Bell a call and see what she charges for a two-line business phone installation and startup. After you recover from the shock, remind yourself that this doesn't include telephones or in-plant wiring.

OK, we're ready to start building parts, but are we ready to start selling? Nope, that takes more cash. We better get someone to sell these super can openers to distributors. No sweat, just sign up some manufacturers' representatives to handle your product.

Manufacturers' representatives are independent salespeople and sales companies that take on several products and sell them for a straight commission. They generally work in one industry, such as pool and spa equipment, or hardware, or whatever. I know a number of reps. One represents components for aircraft fuel systems. He has about six lines, each from a different manufacturer, and has several dozen customers. Another represents products sold in the cabinetmaking business such as glues, hinges, screws, and the like.

Reps don't get paid until you do, so you won't have to put up any cash. Sounds good except for one tiny detail. No good rep will take on a new product line without existing business. They call this pioneering. Only the brand new and very incompetent reps would ever touch a new product without existing business. Sell it yourself? You'll be far too busy handling your day-to-day operation to ever get out and sell. So you'll have to break down and hire yourself a salesperson.

Oops, the salesperson needs a car, insurance, and expenses for a few months. Also he'll need some brochures to sell with. Here's a bare-bones starter kit for your sales effort:

Brochure's photographs and artwork	$ 8,500
Brochure printing (10,000)	2,500
Sign for the building	1,000
Printing—letterhead, invoices	1,000
Salesperson's car	12,000
Salesperson's first month expenses	5,000
Insurance first payments	2,000
	$32,000

Not too bad, you've spent only $133,200 so far and you're in business. Again, this is no ritzy operation. It's just a simple, basic manufacturing business that is able to assemble parts made elsewhere, ship them out the door, and conduct business. We haven't even bought such things as benches for the shop, hand tools, shelving for the inventory, packaging materials, required fire extinguishers, etc., etc. The list could consume several pages. We also haven't hired a secretary to answer the phone, type letters and invoices, and do the books; nor have we hired anyone to work in the shop.

All right. Let's say this has all been done, the tooling is all bought, and you send the salesperson out to sell. Now you can start making money, right? Wrong. The demand for cash hasn't even begun. You'll be spending more than you ever dreamed possible in the months ahead. Let's take a look at why.

Manufacturers are the bankers of industry. When you decide to become a manufacturer, you join the ranks of the largest lending institutions in the country. American industry is based on this fact. The manufacturers invest the most, take the biggest risks, have the most responsibility, and, yes, even finance all of the distribution, dealer, and sales networks beneath them. The manufacturer is responsible for warrantees, breakage, sales and promotional material, and even for supplying inventory to all of the outlets between him and the retail customer, for free.

For sake of example, let's assume the product costs you, all told, 75 cents. This includes the parts, the assembly labor, and the packaging and even has a factor for scrap included. Your wholesale price is $1.75, so you make $1 per unit. Let's also assume that you can run this operation for $15,000 per month counting your, the secretary's, and the salesperson's salary, the phone, rent, utilities, insurance, and so on.

You lay in 10,000 units for inventory and get the salesperson out to sell them in a hurry. He does so, and by the end of the month he's sold all 10,000, and things are looking good for growth. Let's take a look at the books so far:

Sales, 10,000 units × $1.75	**$17,500**
Cost of goods (inventory)	**7,500**
Gross profit	**10,000**
Overhead	**15,000**
Profit ⟨loss⟩	**⟨$5,000⟩**

Not too bad, you lost only five grand. At least that's what the books say. The cash picture is something totally different. Let's take a look at what the real cash situation is by the end of the month.

Cost of inventory	**$ 7,500**
Cost of shipping	**500**
Overhead expenses	**15,000**
Total cash outflow	**23,000**
Income on collections	**0**
Total cash flow	**⟨$23,000⟩**

So you lost $5,000 on paper, but actually you had to put up $23,000 in cash. Why wasn't there any income? Well, distributors don't buy anything COD. They buy it on net 30 day terms. Not only that, but it's a rare distributor who will pay in 30 days. Most will pay in 60, but a few wait 90 days. Do you think you can collect all of your accounts receivable in 30 days? If you could guarantee it, more than one company would be willing to pay you a half million dollars a year in salary. Why would you ever start your own business if you could get a job like that? It is a fact of life for manufacturers and distributors that you have to wait for your money. Remember, you're the manufacturer, so you get to finance your distributor's inventory.

OK, month 1 is over and you plan on doubling for next month. That ought to make you a profit. The salesperson does a good job and meets quota. He sells 20,000 can openers. Great! Let's take another look at the books:

Sales, 20,000 × 1.75	**$35,000**
Cost of goods	15,000
Gross profit	20,000
Overhead	15,000
Profit ⟨loss⟩	**$5,000**

Congratulations. It's only the second month, and you made a profit! You are either very lucky or very clever to do that because it's a rare company that can. But for the sake of example, let's say it happens. Before you start celebrating and throw a party, let's take another peek at the cash flow:

Cost of inventory	**$15,000**
Cost of shipping	1,000
Overhead expenses	15,000
Total cash outflow	**$31,000**
Income, one-third of month 1 sales	5,830
Total cash flow	**⟨$25,170⟩**

That's right. In order to make a $4,000 profit, you will have to dig in your pocket and come up with more than $25,000 in cash. Yes, it's real cash that's required, not promissory notes. Your employees won't work without getting paid, and you have to pay the rent and phone to stay in business, so all of this is real green.

Let's take this little example one month further and double our sales again:

Sales, 40,000 units × 1.75	**$70,000**
Cost of goods	**30,000**
Gross profit	**40,000**
Overhead	**15,000**
Profit ⟨loss⟩	**$25,000**

So it looks as if this little venture is getting somewhere. It appears to make sense. At least it does until we take a look at the cash flow:

Cost of inventory	**$30,000**
Cost of shipping	**2,000**
Overhead expenses	**15,000**
Total cash outflow	**47,000**
Income	
One-third of month 1 sales	**5,830**
One-third of month 2 sales	**11,330**
Total cash flow	**⟨$29,840⟩**

So it seems you still have to put up cash. Another 30 grand! Here's something else I'll bet you never figured on. It's the end of the quarter so Uncle Sam needs his cut. The friendly IRS sends you a bill for 46 percent of your profits, and that's on the book profit, not the cash profit. Your profits so far are a $5,000 loss in month 1, a $5,000 profit in month 2, and a

$25,000 profit in month 3, which means a $25,000 profit. Forty six percent is $11,500. Again you get to dip into your pocket.

Let's sum up the cash expenses for starting this little manufacturing business where you get everything made outside, bring it into a little warehouse where just a few people work, and sell it for a good profit:

Startup expenses	**$133,200**
Month 1 cash requirements	23,000
Month 2 cash requirements	25,170
Month 3 cash requirements	29,840
Tax bill	11,500
Total cash requirements	**$222,710**

It sounds ridiculous, but it's true. You'd need nearly a quarter of a million dollars just to start up this little business and run it for three months. It's profitable besides! You might say that you'd scrimp and cut corners, but the truth is you couldn't make a big enough dent in this huge cash requirement to really make any difference. If you work this example out a few more months, you'll find out that the cash demand doesn't get smaller; in fact, it gets bigger. How can that be?

One of the amazing things about a manufacturing business is that as long as it grows, it consumes cash. Not only that, but its growth is limited by how much capitalization cash is available. Companies actually have to turn down orders because they do not have sufficient cash available to fill them. The only time the company quits consuming cash is when it quits growing, or at least the growth slows down to less than the after-tax profit percentage. This means that most manufacturing businesses cannot grow more than about 10 percent per year without consuming big bucks.

What does all this mean to the average person with a new product idea who wants to make a buck? Forget making the product yourself. Forget having it made and selling it to the world. By having these ideas, you are a creative person. That is good. Nothing will stifle your creativity faster than struggling through the daily rigors of managing a manufacturing business. The headaches are huge. To many it is their profession. Leave it to the manufacturing experts. You can do far better by exercising your creative juices and coming up with another new product. I've watched too many people go rapidly broke trying to build their products themselves and not heeding this advice.

For you, the creative genius, the best route is to sell your product to a manufacturer who will take care of all those things for you and just write you a check every few months for your product ideas and development. Meanwhile you'll be free to develop yet another product while the royalties roll in. If you develop two or three winners, you can quit working altogether while the royalty checks keep coming. Just have them forwarded to your mail stop in Tahiti.

7

How to

Sell

and

Profit

from

Your

Product

If you've been following my steps through this book, you're probably pretty tired of putting lots of time, effort, and money into your product and are ready to start making a few bucks off of it. The process of selling a developed product is the step at which most would-be inventors make glaring errors and end up throwing up their hands in disgust. Actually, it is not at all difficult if you follow the steps outlined in this chapter.

If you have a well-developed product ready to introduce into the marketplace, there are literally thousands of companies out there eager to make it and pay you handsomely. The only real difficulty is finding them and presenting your product to them.

While writers have reference books listing all the publishers of books and magazines looking for written material, and shoppers have the yellow pages of all the stores looking to sell them something, unfortunately, new product developers do not have a reference book listing all the companies looking for new products, so it takes some research to figure out who may be looking for a product similar to yours.

Locating a Suitable Manufacturer

Your basic research begins with trade journals. Trade journals are magazines for specific industries. These are not magazines you can find on any newsstand; they are circulated among the manufacturers, distributors, and dealers of specific product groups. For instance, there are four or five big trade journals for the pool and spa industry, several for the hardware industry, plumbing industry, toy industry, gift industry, kitchen utensil industry, even Christmas tree ornament and fireworks industries. There are trade journals for practically every conceivable group of businesses that you can imagine.

Your first step in researching a possible manufacturer is to thoroughly read every trade journal you can lay your hands on for that industry. You can locate the names of these journals in any local library. Look for a reference book of periodicals. They all have trade journal sections. Remember, you're not looking for consumer publications, but publications only for the trade. For example, if you invented a new accessory for cameras and are looking for a suitable manufacturer, you would not be interested in a magazine such as *Popular Photography*, which you can buy on the newsstand, but should get a photographic equipment manufacturers' association trade journal.

Trade journals offer a unique view into the business of a particular industry. People, products, new technology, and opportunities are presented in a totally different way in trade journals than in commercial magazines. If the Duluth Fork Company was recently purchased by Minnesota Silverware Enterprises in an attempt to increase their Northwest market share, you'd never expect to find the transaction in the *Los Angeles Times* business section; nor would you ever read about it in *Better Homes and Gardens*. You would, however, read about it in one of the silverware trade journals.

If you study the trade journals, you can find out all sorts of interesting nuggets of information about that particular industry. You'll know which people make the decisions, which new products are coming out, which companies are aggressively growing, and in short, everything you could ever want to know about that industry.

If you just developed a new type of eating utensil, the fact that Minnesota Silverware Enterprises is out there buying up companies may be vital to the sales of your product. It is just this kind of information that will make you rich. Lack of it will make you frustrated.

Here's the good part. Most trade journals are free to members of the trade. All you have to do is develop a list from your local library, write to them and ask for the last few months' issues along with an application for subscription. You should also explain that you are involved in the trade as a new product consultant, or some other suitable job description. Don't lie and say you have a huge company making some product for the industry. Most industries are very small. Everyone knows everyone else, and professional reputations stick for years, even if undeserved.

The thing to look for when reading a trade journal is smaller but aggressively growing manufacturers. You can get this from the industry news sections and maybe from the advertisements. Avoid large, conservative, established companies. They often get set in their ways. Keep a close lookout for shakeups, that is, companies being bought and sold, new divisions being started up, and new managers in key positions. These signals can give valuable clues to the basic policies of the companies and the general satisfaction of the owners or stockholders.

Compile a list of the companies that make products similar to but not directly competing with the one you've developed. If, for example, you've developed a new type of bicycle rack for cars, you may have a lot more luck selling your product to a ski rack maker who wants to expand his product line than you would selling to an existing bicycle rack manufacturer.

Once you get a list of potential manufacturers, you need to compile some information about the companies. There are three excellent sources of information you'll want to consult. The first is the *Thomas Register.* The *Thomas Register* is an eighteen- or twenty-volume set (it changes every year) listing all of the manufacturers of everything you can imagine. The register, available in any library, is made up of three parts. The first is categorized by products, with lists of companies that make each product.

The second part of the *Thomas Register* is called Company Profiles and will give you an idea of the size of the company, what they make, and where they are. The third part contains catalogs of the products of some of the companies. Companies have to pay for catalog listings so not all of them appear there. Remember the *Thomas Register* is for company product information.

To get financial information about a company, consult Dun & Bradstreet. They list most every corporation in their *Million Dollar Directory*, compiled by Dun's Marketing Services, Inc. All of the financial records are given for publicly held companies (ones whose stock is traded on the open market). Information about the privately held companies is more sketchy, but you'll be able to get some idea of their size and condition by the number of employees, sales, and other published data.

To find out who the people are in a company, a reference book called a manufacturers guide for each state is printed. You may have some difficulty locating a guide for companies in states other than the one you live in, but a helpful reference librarian in a big city library can usually find the information for you.

You should amass all of the information you can about every potential manufacturer of your product. Send away for the company's literature. Talk to local distributors of their products. Talk to their dealers or customers if you can. Get everything you possibly can about the companies.

What you are looking for is companies that:

1. Are smaller but growing

2. Have young, aggressive management

3. Are well capitalized (divisions of larger companies or with a reputation for having lots of cash)

You could even come up with a rating system for the companies best suited for your product. Believe me, trying to get into General Electric or Mattel Toys will be difficult to the extreme, unless you have a reputation for coming up with winners. If that is the case, the bigger companies will seek you out. However, getting into Southern Idaho Toy Company, recently purchased by a Chicago-based conglomerate of plastics products manufacturers, will be a breeze compared to Mattel.

How Manufacturers Work

If you're going to make any headway with a manufacturer, you have to know a little about how they are organized. Generally there's the big boss. His title is president or general manager or sometimes something else. Under him are three key "line" people and sometimes a few "staff" people. Line people are people directly involved with the decision making and operations of the business, while staff people are more advisory types. The president's main responsibility is "P and L," or profit and loss. The big boss's primary responsibility is economic. His only real concern is to run the company so it makes money. Every other duty is secondary and every decision is based on the effect that decision has on the profit and loss.

The three line people below the president are the manufacturing manager, the engineering manager, and the sales manager. These people sometimes have slightly different titles, but their main duties are directly connected with the product and business of the company. All other executives are in staff or advisory positions. Executives with titles such as director of market research, corporate controller, director of R & D, and all of the other fancy names you might hear are staff personnel. Their sphere of influence is advisory. Few actual decisions about the direction of the company come from these executives. Don't get me wrong—I'm not belittling their duties or effectiveness, just pointing out that these are not the right people to sell your idea to.

Let's take a look at the line managers. The manufacturing manager is primarily concerned with getting the product out the door. He wants no changes and no problems. His job is to ship as much finished product as he's able every day.

Next is the engineering manager. His primary responsibility is to establish and control the product the company makes. He has to be concerned with the fit, function, appearance, and manufacturability of the product.

The last member of this triad is the sales manager. His

duty is to sell the product. His mandate is to produce orders to ship the product to. His job hinges on growth of the organization.

So we have the essentials of any manufacturing enterprise: establish the product, sell the product, and make the product. Everything else in a manufacturing company is secondary. Having been a member of this group in several manufacturing companies, I can assure you that these three people seldom, if ever, agree. They may be buddies and go to happy hour together after work, but in the course of business, they're bitter enemies. This is almost by definition. The engineering manager wants to make some product improvement changes, but the manufacturing manager just got his people trained to making it the old way. The manufacturing manager wants to simplify the product to make it easier to produce, but the sales manager wants some more features so it's easier to sell. On and on it goes in heated conference room debates.

Well, after this dissertation of how a manufacturer works, you've probably guessed that the person you want to talk to about your new product is the engineering manager. If you said yes, you're wrong. The engineering manager usually suffers from what salespeople call the "NIH syndrome," which means Not Invented Here. In other words, if they didn't invent it, it's no good. Engineering people become so specialized and confident in their knowledge of their particular product that they often become closed to new ideas infringing on their turf. It's a protective response.

The correct answer to the question of who you want to present your product to is the sales manager. His job is to make the company grow. He's the one whose job is on the line if the volume doesn't support the overhead. He's also the one looking for that slight edge to give him the advantage over his competitors. The sales manager is the only one you need to present your product to. His decision and assessment will make or break you. All business begins with the order, and the sales manager is the one to get that order. Every one else in the company may love the product, but unless the sales manager is sold, you don't have a chance.

There are some other advantages to selling to the sales manager. One is that salespeople are generally the easiest people to sell to. They usually listen carefully and are masters at balancing the pros and cons of a decision. Sales-oriented people are accustomed to making quick decisions, as opposed to engineering-oriented people, who need to analyze and reanalyze and check every decision.

The sales manager is far more accessible than other managers, too. While the president, manufacturing manager, and engineering manager are so consumed with internal duties that they are isolated behind secretaries and several tiers of phone call screeners or are off in meetings, the sales manager is the communicator from the company to the outside world. It's his job to receive input for the company. It's far easier to get to see him than anyone else at that management level.

All of this means that you must first make an appointment with the sales manager. If you did your homework on the company, you know his name. If you couldn't get his name, call the company and ask the receptionist. This is done all the time. Just call and ask who he is, then ask to speak to him.

When you have him on the line, don't get cute by asking how he'd like to be a part of the greatest product the world has ever known or would he like to make a million dollars or any of these types of trick questions. He's a professional and busy besides. Just introduce yourself, explain that you have a new product you think would fit in his line quite nicely, and ask for an appointment.

Don't get into any verbose descriptions of your product, either. You want to describe in very sketchy terms what the product is: "I've developed a new type of _____ that I think would fit with your products well, and I'd like to discuss it with you. Can we set up a meeting so I can show it to you?" You want to introduce a little mystery so he's eager to see it. You also want to let him mull it over in his mind for a few days so he can do some speculating on the possibilities, so make sure you set the date for the meeting at least four or five days ahead. Don't ask if you can drop by that afternoon or the next day.

If he's uninterested or flatly turns you down, don't worry about it. Just thank him for his time and go on to the next company on your list of candidates. Certainly his turndown is no reflection on you or your product. You'll never know, nor should you care, why he turned you down. It's possible they're working on a new product such as yours. It's possible the boss just chastised him for going on wild goose chases. Rejection should not be taken as a personal affront. In fact, many people, myself included, enjoy the rejection, because they know it brings them that much closer to success. It also gives you a little incentive when your product does make it and is selling like hotcakes. Just think about that poor slob wishing he'd hooked up with you first.

I've gone through these steps at least a dozen times and have never had to call more than three companies to get an appointment. Many never panned out, but some did. It's a numbers game. Just keep at it, and soon you'll have an appointment to show your product.

Before Your First Meeting

Well, you've made your appointment with the sales manager of XYZ Widgets Company, and you're raring to get in to see him. Before you fly in there with all of your notes, drawings, and prototypes, there are some important numbers you must have to interest him.

You'll need to calculate a cost of the product. In Appendix A, I've put together some very rough numbers for various manufacturing processes. Use these only as a general guide. It would be much better to get your product estimated by a job shop specializing in the processes you need. If you have machined parts, send drawings of the machined components out to some machine shops. Similarly, plastic injection molded components should be sent out to injection molding shops for a quote. You'll have to specify some materials and quantities for these shops to give you an honest price.

You'll need to get prices on every component, even the screws and washers. The *Thomas Register* is a good source of manufacturing job shops (companies that make parts to a customer's specification). Another good tool is the yellow pages.

After you get all of the piece part prices, you'll need to figure out some value for assembly. Use about $25 per hour as a shop rate and guesstimate how many of your product a trained assembly person could put together in an hour. If your new product is electronic or something that lends itself well to mass automated production, use a higher figure, like $50 per hour.

Another cost I'll bet you've never thought of is packaging. Don't underestimate the cost of packaging. For many products the packaging is more expensive than the product. Let me give you an example: salt costs about $20 per ton in bulk. That means 1 cent per pound. Believe me the little round salt boxes with the metal pour spouts that you find in the supermarkets cost far more than a penny. All sorts of products fall in this category—soft drinks, shampoo, toothpaste, makeup, etc.

You must also include the shipping cartons. Look at toothpaste, for example. The toothpaste is in a tube, which is in a box. Twelve individual boxes are wrapped in a carton, and twelve cartons are packaged in a case. That means you'll need tubes, tube boxes, carton boxes, and case boxes. Each of those will have to be printed. There are many packaging companies that will look at your product and design and quote on appropriate packaging. Check your yellow pages.

Now you know the costs. To get your approximate wholesale price, multiply this cost by five. That's right, five times the hard costs. If that shakes you up, go back and reread Chapter 6. This is a typical markup for manufacturers. If that throws your product price way out of the ballpark, don't think the manufacturer will take less profit, you'll just have to find a cheaper way to make it because they won't touch it for a low margin.

All right, you have your tooling costs, your product costs, and your wholesale price. The wholesale price minus the product cost is your gross profit. Divide the tooling cost by the gross profit and you'll get the break-even quantity to pay for the tooling.

Before you show up at your first meeting, you should have the following numbers:

1. **Total costs broken out by parts, assembly, and packaging**

2. **Approximate tooling costs**

3. **Approximate wholesale price (5 × cost)**

4. **Gross profit per unit (wholesale price − cost)**

5. **Tooling break-even quantity (tooling ÷ gross profit)**

There are actually some other costs that the manufacturer will add such as advertising, sales, and overhead, but don't presume to know or guess these. The manufacturer will be painfully aware of what these will cost him. You don't have to remind him.

You'll also need plenty of copies of your secrecy agreement and your prototype before your first meeting. You should also go over the features and benefits of your new product until you have them down cold.

There are differences between features and benefits, and you shouldn't confuse them. A feature is something unique about the product. A benefit, on the other hand, is why you'd want to use the product rather than something else. For instance, if you developed a bicycle storage rack for garages, a feature might be that it's lightweight and can be moved around easily, while a benefit might be that it makes someone's garage neater. (Don't laugh about my bicycle rack example—I invented one of those once. It didn't sell, but then again, I offered it to only one manufacturer, and he wasn't right for the product. One of these days I'll have to resurrect that project.)

Your First Meeting

The big day has arrived, and it's time to show up for your first meeting. You should dress nicely in business attire. It may be grossly unfair, but people do judge you on your looks, so if you look the part of a successful businessperson, the people you're dealing with will assume you are. There's one saving grace, though—inventors and scientist types are allowed to be a little eccentric. If you show up in blue jeans and unshaven, though, no one will take you seriously.

Stash a bunch of papers and notebooks in your briefcase. Put in a few business cards, even if they're only from your florist or mechanic. I've noticed that whenever anyone opens a briefcase, everyone in the room glances inside. I guess they hope to get a little insight into the person by looking in his briefcase. You could include a calculator and a vernier caliper, too, to lend some technical credibility and make it look as if you actually use the briefcase. I've noticed that when someone shows up with a brand new briefcase with only one file inside or just a few notes, people become suspicious.

When you get into the sales manager's office, give him a warm smile and handshake and take a good look around. You can learn much about a person from the looks of his office. Is he organized? What are his interests or hobbies? I closed one deal by having an hour-long discussion about the manager's daughter's soccer team. I asked about it, then sat back and listened. By the end, he and I were like old school buddies. The discussion of my new product took only 30 minutes after that, and he was sold. It is very acceptable and sometimes even expected to engage in a little small talk before getting down to business.

When it's time to discuss your product, the sales manager may speak to you alone or may lead you off to a conference room while he gets a few more people to listen in. You should hope to get him alone, but if he wants to bring in the engineering and finance people, it probably means they are quite serious about expanding and introducing a new product. Whether you are talking to him alone or in a conference room full of bodies, keep in mind that it's still the sales manager you want to sell.

Before even opening your mouth, you should distribute copies of your secrecy agreement and have everyone in the room sign them. If they are reluctant, they probably plan to steal it from you. Most people won't be at all reluctant. I've even seen companies that have their own secrecy agreement forms for people who don't have their own. Be suspicious if they won't sign, and don't show them the product or discuss it technicallly. You could discuss it in very general terms, but I'd be inclined to thank them for their time and leave.

Once you have the agreements signed, pull out your prototype, set it on the table if possible, and give them a little rundown of the features and benefits. Be prepared for some type of demonstration if at all possible. Obviously, some products don't lend themselves to demonstration, but often the company will have some sort of test facility to try out the prototype.

After you've discussed the product and demonstrated it, go ahead and bring up the economic feasibility (cost to tool, cost, sell price, break-even quantity, etc.). This will demonstrate to them that you know what you're doing and you've thought it out.

The next step is to shut up and listen. Pay close attention to what's being said. You'll probably have to field questions as well, but keep your answers short and to the point. At this stage you should try to get some idea of what they think. How does each of the decision-making people feel about the product? Who likes it and who doesn't? What are the objections? Which guy in the room commands the most attention? Store these little impressions away as a squirrel stores nuts for winter. Each of these bits of data will become vitally important when it comes down to negotiating.

Don't expect a decision to be made on the spot. Generally either these meetings end with a rejection, or they are left in the air, a big maybe. Also don't discuss what you want for the product in the way of buyouts or royalties. If you're asked directly, be evasive. Tell them that you're open for negotiation, or that you'll have to work it out. Do not talk money for you at the first meeting. First get them to want the product, then you can talk about how much it will cost.

The three biggest mistakes you can make in the first meeting are:

1. Talking too much about the product and defending it when you hear objections

2. Trying to get a commitment

3. Bringing up how much you want for the product

All of these points are much better left to future meetings. The first meeting is a sales meeting. Get that company to want the product!

Unless the people at the first meeting flatly turn you down or your prototype failed miserably, get a commitment on the next step. Even if it is only a commitment for a phone call in a week, you should leave the meeting with a definite plan of action. You should know what the next step is. Perhaps they'll ask you to build three more for testing, or, more likely, they will tell you that they have to discuss it and think about it for a while. Don't immediately think this is a rejection. It most emphatically is not. The amount of decision making, numbers projections, customer surveys for reactions, and manufacturability studies that have to go on before a decision on a new product is made is staggering. There are a number of weeks of work that have to go into it, even if they love the product.

Don't forget Mr. Murphy, either. I once developed a new type of self-indicating bolt that would electronically sense how tight it was. The bolt worked perfectly in my lab on the morning just before my first meeting. In the meeting, all of the people were very excited about the prospects for this new product. We walked out to their test lab for a demonstration on their test machine. Wouldn't you know it, I had slightly overstressed the bolt on my last test so it wouldn't work at all on their machine. After all of the rhetoric on how my new bolt would revolutionize the industry, all I got was thoroughly embarrassed. The moral of the story—take several prototypes if you possibly can. I had only one with me.

After the first meeting, wait for the call back. If you have not heard in two weeks or so, call them and ask if you can offer any further information or answer any other questions.

You'll get either put off or rejected. If you get put off to a later date, ask when you should call again. That's an excellent way to find out how the product went over. Often, you'll hear what's happening or who is unsure and needs more time.

Prices, Terms, Royalties

Just about everyone who ever invented anything thinks his product is worth a million dollars. Maybe it is, but unless you just invented a time machine, I'll guarantee you won't find a manufacturer who will give you a million dollars up front. It just isn't done that way. If you believe in the product, you should be willing to accept royalties.

A royalty is a payment to you for a product based on how successful it is. It is usually a percentage of the wholesale price. Sometimes, though, it is a flat rate per unit. Don't expect this to be a huge number, either. It runs from 2 to 10 percent of the wholesale price, with the majority in the 3 to 5 percent bracket. Consumer products with huge dollar potential have lower royalties, whereas specialty products are sometimes higher. Don't scoff at these seemingly low numbers. A consumer product can easily get to the $10 million a year sales range. Two percent of $10 million is $200,000, nothing to turn your nose up at, especially since you won't have to do anything once the manufacturer takes over.

I currently get 3 percent royalties for a high-dollar consumer recreation product, 5 percent for an industrial component product, and 10 percent for two specialty aerospace products. On both the consumer product and the industrial product, I received reimbursement for development costs. One is a five-year term and the other ten. On the two aerospace products, I had to develop the products on my own, but the term is essentially forever.

It is acceptable and reasonable to ask for development expense reimbursement from the manufacturer. Don't go overboard, though. Usually the development costs include all of your cash outlay in getting the product built, all extra expenses such as travel and testing, plus all your time at some reasonable rate.

The typical license contract runs from $5,000 to $50,000 down and 3 to 5 percent for five or ten years. All these numbers are rough, and they depend on the complexity, the selling price, the amount of work that has gone in, the amount yet to go in, and the sales potential of the product. All license contracts are negotiable. They're free-form, so they can be written up any way you and the manufacturer can agree. There are no set rules.

Don't let greed get in your way when negotiating these contracts. Your product is not worth anything unless someone manufactures and profits from it. No one in the world will give you a dollar royalty for a $3 wholesale item. Remember, most manufacturers make about a 15 percent profit selling goods that they mark up five times over direct costs, so no one will be getting rich by giving you 5 percent. Look at it this way—with five or six winner products out there you could easily be making a six-figure income on royalties for ten years without ever doing any more work. Your best bet is to get your first one sold, no matter how good or bad a deal you end up with. It establishes you. It gives you the incentive to go out and do it again. Once you have five or six winners, then it's time to get hard-core about your contract negotiations. But you must have a track record first.

After all of the negotiations are through, make sure you get all you want in an official written contract. The chances are almost 100 to 1 that the manufacturing company will prepare a contract. Make *sure* the manufacturer doesn't try to pressure you to sign the contract on the spot. That's a sure sign of a scam. You must take the contract with you and consult a qualified attorney.

No contracts are "standard." Anyone who tells you that is out-and-out lying. Each one is unique and depends on the terms, conditions, stipulations, and reimbursements negotiated. Every contract can be changed before it is signed. After it's signed, you're stuck with it—that is, unless you can prove you were either willfully misled or unfairly pressured. Believe me, either of those two cases is mighty tough to prove.

If the company asks you to draw up the contract, go straight to the sharpest business attorney you can find. I might spend twice as much as is needed to get as high-powered an attorney as I could find if the product had some major potential. It would be money well spent. Be sure you have some method of auditing the amount, either in number or in dollars, of your product that the company makes and sells. All the contracts in the world won't help you unless you have a way to verify the sales of your product. Trust, but verify.

Persistence Pays Off

If you have a new product and have developed it, you have a valuable thing. It's worth something. Someone out there wants it. Keep that in mind, even if you get turned down dozens of times. There is a company out there just wishing you'd walk through the door and save the day. You have to keep looking. Persistence is a must. You'll never make it in the inventing business unless you are doggedly persistent. Remember, rejections just mean that you are one step closer to finding that manufacturer who can't live without your product.

Richard Bach, the author of *Jonathan Livingston Seagull*, was turned down by book publishers more than thirty times before he found a publisher willing to do his book. Once they did, it became a best-seller and was reprinted many times in several languages. Imagine all those publishers out there who turned it down and are now sick with envy about how well it did and wishing they had decided in favor of publishing it.

Manufacturers are the same way. Many turn down the greatest thing that could happen to them because of lack of interest, poor speculation, or any one of millions of other reasons. Can you imagine the inventor of the Hula-Hoop going from conference room to conference room demonstrating his invention in front of stern-faced executives. The scene is hilarious, yet the inventor eventually did find someone to make it and became quite wealthy in the process.

Keep at it, keep at it, and keep at it are the three most important bits of advice I can think of to give to the aspiring inventors of the world. Persistence is the quality that separates the ones who go out and make it from the multitudes who only dream of making it.

Appendixes

Appendixes

Manufacturing Processes

In this section, I describe different manufacturing processes for the new product developer. These descriptions are intended only as a guide. Economies change, and products, tooling, and process costs change right along with them. This is not to say that they always go up. Oftentimes bargains can be found for some processes and equipment.

Regional economies can greatly affect the local cost of tooling. For instance, in the early 1980s when oil prices dropped drastically, the local economy of Houston, Texas, was in serious trouble. Much of the local business is geared to the oil industry; when prices drop, so do the support businesses for the industry. If you had developed a new product that uses some of the same manufacturing processes as are used to make oil drilling tools, you could find a great many bargains. In fact, you probably could have had an auction to see who would have done it the cheapest.

On the other side of the coin, in the late 1980s in southern California, the precision manufacturing shops were totally glutted by all of the aerospace work. In this type of economy, you can hardly find a shop willing to bid on making your product. If you do, it is usually expensive, with a long lead time.

Other regional differences influence the type and cost of tooling and manufacturing. For instance, I once did some shopping around for a plastic injection molding tool. It was for a part that looked roughly like a laundry basket, except half the size. In southern California most of the injection molding is for electrical connectors and medical equipment: small, precision parts. The best price I could find for the tool was $55,000.

Instead of going for it, I jumped on a plane and made a tour of the Midwest—Chicago, Detroit, and Cleveland—looking for a mold maker with better prices. I found that in Chicago, many of the molders serviced the appliance industry and were accustomed to medium-sized and large round parts. The best price there was $31,500. So with a trip that cost a few hundred dollars, I managed to save $23,500! Not bad for three days' work.

Don't be afraid to shop around all over the country for bargains. Much of it can be done by mail, fax, and phone. Consult the *Thomas Register*, available in any library, for tooling and manufacturing services. They list these services by category and by city.

This listing will help you decide how your new product should be built. Products can be built in many ways. Deciding which processes to use depends on how many are being built. One manufacturer of a product I developed began building the plastic components of the product by the plastic machining process. This was higher in cost than molding, but it didn't require expensive tooling. Once the volume got up to where it made sense, he switched over to plastic injection molding, a process that is far less expensive for each part made, but requires a substantial up-front expense for tooling.

One thing to consider when going through this list is that most companies that perform these processes have a minimum cost for sampling the tooling and getting you a proto-type. Recently I needed six parts to test. The plastic molding shop charged a standard $300 sampling fee. I ended up with about fifty parts. You cannot get just one part made if tooling and setup time is involved. One part would cost exactly the same as fifty. It turned out that on this part adjustments were needed which required some tooling modifications. The second sampling cost the same as the first—$300. So to get six usable parts, I had to invest not only in mold tooling, but also $600. That's $100 per part, a bit expensive for a plastic part barely one inch across, but that's what it takes to prototype right. Once this part goes into production, it will cost less than a nickel.

Again, the dollar amounts are very general and are intended only to give the new product developer an idea of the comparative amounts that manufactured products cost.

Metal Processes

CASTING

Casting is one of the oldest processes for forming metal. In the casting process, metal is heated to the melting point then poured into a mold the shape of the desired end product. Just about any metal can be cast. There are a number of types of casting. The simplest is sand casting. Automobile engine blocks, water valve bodies, hibachis, and similar things are sand cast. Sand castings leave a rough finish on the parts which later has to be machined smooth for flanges, holes, and fitups.

Typically, the tooling for sand casting is inexpensive, in the $200 to $2,000 range. Complex shapes can be made readily. The casting for a 12 × 12 inch hibachi would run in the $600 range for tooling and from 75 cents to $1.50 for the parts, depending on quantity.

Another popularly used casting method is called investment casting. This uses a lost wax method to create the cavity with a plaster of paris mold. More precisely finished parts can be made in this way. Lock parts, gun parts, and keys are often investment cast. Tooling costs more than for sand casting, running in the $1,000 to $10,000 range. For a key blank, the tooling will run about $1,200 with a piece part price in the 10- to 30-cent range, depending on quantity.

One of the newest casting methods is die casting. In the die-casting process, a hardened steel mold is made and molten metal injected into the tool under pressure. This process is only for such low-melting-point metals as aluminum and zinc. Very accurate, complex shapes can be made. Such things as automobile name plates, carburetors, and Matchbox car bodies are die cast. The tooling is expensive, but the piece part costs are low. It is suitable for high-volume production only. Typical tooling cost for a 2-inch-long model car body is about $8,000 to $10,000, with the per part cost about 5 cents. An automobile carburetor body costs upwards of $30,000 in tooling with about a $1 part cost.

MACHINING

Machining is the traditional method for forming shapes of metal. Machining begins with a piece of metal larger than what is needed, then metal is cut away with specialized machines until the desired size and shape are attained. Most gears, shafts, fittings, threads, pulleys, valve internal parts, and automotive internal parts are machined.

Machining is very volume-sensitive but has practically no tooling costs associated with it. A brass sprinkler jet, for instance, would cost in excess of $20 for one, about $2 each for 100, and probably 20 cents each for 10,000 parts.

Many parts that normally could be made with other methods in production have to be machined to make prototypes. Take, for instance, the dead-bolt piece on a lock. To make two or three for prototypes would probably cost $2,000 by the investment casting process, yet they could be machined for about $150. For 10,000, however, investment casting would probably cost in the $12,000 range including tooling, but machining would cost $30,000.

STAMPING AND FORMING

Stamping is an economical method to make many parts. Electrical connectors, pots and pans, automobile grilles and body panels, appliance parts, latches, switch plates, pen clips, sinks, and a host of other common metal parts are stamped. As for many processes, the tooling is expensive, but the price per part is low. An 8-inch aluminum frying pan shell costs in the $5,000 range to tool up and probably $1.25 each in quantity. A complex shape like the clip on a pen might cost $10,000 to tool and about 1 cent per part in big quantities.

FORGING

The forging process is used when the finished part must have high strength. Most tools such as wrenches and screwdrivers are forged. Some other forged items are automobile crankshafts, camshafts, aircraft landing gear, and hammer heads. Forging is expensive to tool. A commercial aircraft landing gear forging tool might cost half a million dollars. Something like an open-end wrench tool would run in the $10,000 range with piece part costs in the 50-cent range for a smaller wrench. Most forged parts require finish machining.

EXTRUSION

Extrusion is used when a long thin part with a constant cross section is needed such as aluminum window frames, metal tubing, and many metal trim pieces. Extrusions are relatively inexpensive to tool up and are generally sold by the pound. A typical extrusion die runs about $300. An aluminum window frame extrusion costs about $2 per pound and may have about 5 feet of extrusion per pound.

OTHER METAL FABRICATION

Many other processes are used for making metal parts. These are generally lumped together into the term "fabrication." This includes the cutting and bending of sheet metal and bending and welding of metal assemblies such as fireplace grates, machinery frames, automobile jacks, and the like. Shops that do this type of work are usually all listed in the industrial directories under the heading of metal fabricators. The costs and tooling for these items are so diverse that to go into detail would just introduce confusion.

Plastic Processes

PLASTIC INJECTION MOLDING

Injection molding is the most used process for the high-volume manufacture of plastics. Telephone shells, toys, plastic handles for appliances, computer casings, new car dashboards, radio cases, knife handles, plastic utensils, knobs, curtain rings, pens, Frisbees, disposable razors, electrical components, newer hub-caps, steering wheels, and even toilet seats are just some of the myriad of injection-molded products made today. Just forty years ago, none of these things was made by this process.

Injection molding is fairly expensive from a tooling stand-point, but is one of the least expensive methods to manufac-ture anything in high volume. No secondary operations are necessary in most injection molding operations. If you've ever built a plastic model car or airplane, you know that the parts are all attached to a "tree," which is exactly how they come out of the mold. The tree is actually the channel that the molten plastic took to fill the mold.

One convenient feature of plastic injection molding is that many parts can be made at once. I once saw an injection-molding tool used to make cocktail toothpicks (the kind that olives come on). The tool made more than 200 parts at once. The machine time governs the cost of a part, and it is exactly the same whether you're making one part or many at once. For instance, to make one toothpick takes 10 seconds. At a penny per second ($36 per hour), that's 10 cents per toothpick (plus material). If in that same 10 seconds you made 200 parts, it would only cost .05 cents each (plus material). The material to make one toothpick is exactly the same in either case.

A single-cavity tool to make a small plastic bowl, for instance, would cost about $7,500; a four-cavity tool (which makes four bowls at a time) would cost about $20,000. The piece part price for the single-cavity tool would be about 20 cents and for the four-cavity tool about 7 cents. The tool to make a telephone housing might run $100,000 with a 75-cent part price. One of the most expensive tools I've ever seen makes the front grilles of the new Corvettes. It cost nearly half a million dollars. I suspect the parts cost in the $10 dollar range.

PLASTIC MACHINING

Plastics can be machined just as metals are. The same machinery and equipment are used as for metal machining. When selecting a plastic machine shop, however, make sure that the shop is familiar and has had experience with plastics. There are some different techniques and tool shapes required for plastics, so stick with experienced shops. The costs run about the same as for metals.

PLASTIC BLOW MOLDING

Blow molding is the process used to make bottles and other thin-wall, hollow parts of plastics. To make a typical bottle, the plastic is melted, then collected on a hollow tube around which a steel mold is closed. Air blown into the center of the molten plastic causes it to swell up like a balloon until it contacts the sides of the mold and cools and the mold is opened forming the complete bottle.

To tool up for blow molding is in the $2,000 to $5,000 range for a single-cavity mold with a piece part cost in the 20-cent range. Like injection molding, multiple-cavity tooling can be made for high-volume production.

PLASTIC EXTRUSION

Plastics can be extruded just like metals to make long, thin, constant cross section parts. Hoses, tubing, trim strips, and strips of plastic are generally extruded. Tooling is inexpensive, in the $100 to $500 range, and the part cost is usually by the pound.

Rubber Processing

Rubber is processed into finished shapes by the compression molding process. Small parts such as rubber feet, crutch tips, electrical connector covers, grommets, and similar types of parts are compressed in a mold under heat until the rubber cures. Rubber molds are inexpensive, in the $100 to $1,000 range depending on the number of cavities. The piece part price is usually inexpensive as well, running about 3 to 5 cents for a small part made in a multicavity mold.

Electrical and Electronic Processes

PRINTED CIRCUIT BOARDS

Printed circuit boards are the standard in electrical and electronic construction today. No devices on the market are made using discrete wiring. A printed circuit board must first be designed, then artwork made. The artwork is then photo-transferred to the copper-clad board. Acid that affects only the areas exposed to light eats away the copper where it is unwanted, forming the circuits. The board is then drilled to complete the operation. There are kits available in just about every electronic parts store if you want to try your hand at making one or two boards for prototypes.

For a production environment, the initial design and artwork stage runs in the $1,000 to $10,000 range depending on complexity, and the piece part price from about 50 cents to several dollars, again depending on size and complexity.

BOARD STUFFING AND SOLDERING

The components of a circuit must be assembled to the board and soldered in place. For a few, they can be hand-stuffed and soldered, but for any high-volume production, automated machines are used. There are companies that specialize in this process and charge according to volume and complexity. Because every circuit is unique, speculation as to cost is more confusing than helpful. Obtain quotes from local vendors.

Assembly Processes

Mechanical and electrical assembly shops abound. Many are connected to charitable organizations such as disabled associations, career retraining groups, and assistance groups for the physically and emotionally handicapped. Often these organizations are subsidized by government or church groups and so are a bargain. Consult your local industrial telephone directories for these organizations in your area.

Filing a Disclosure Document

A disclosure document is not a patent. It doesn't give you any rights that a patent holder has; it really does very little except establish you as someone who holds prior art. It also allows you truthfully to say that you've filed with the Patent Office when selling your product.

If someone violates your secrecy agreement and comes out with the product you first introduced to him without paying you a royalty, you can sue for breach of contract. Suppose also that the company insists you never told them of the technology—you're out of luck unless you can prove you did indeed develop the technology prior to your discussion. Your laboratory notebook is some proof, but that can be made up after the fact and predated. Your best proof is a filed disclosure document. The U.S. goverment is holding that document, and they've issued you a number. What better proof is there?

A disclosure document must contain certain information to be a valid document. It should have:

1. Title page

   On the title page put the name of the invention or new product, your name, the date, your signature, and the words "Disclosure Document."

2. Introduction

The introduction simply states that this is a disclosure document of a certain product. I also like to include a sort of attestation that I believe this product to be unique, that I designed or developed it, and that I intend to pursue a patent on the product.

3. History or Background

In this section you describe why the product has been developed. In it you can also state what similar products are out there and why your product is different or unique.

4. Description of the Product

In this section, you describe only the product, not how it's used. Merely describe the physical features as they appear on the product. See the sample disclosure document that follows.

5. Drawing or Sketch of the Product

This should be done in ink, on 8½ × 11 paper. It should not be in engineering drafting style, but more a technical illustration of the device. A patent-type drawing is preferred but rather than going to the effort and expense of hiring a patent draftsman, just do the best you can with the talent you can find. For a disclosure document, a drawing that shows the pertinent information that makes your product unique is all that is really needed.

Once your disclosure document is complete, send two copies to the Patent Office with a check or money order for $6. Send the document via certified mail, with a return receipt requested, to:

U.S. Department of Commerce
Commissioner of Patents and Trademarks
Disclosure Document Program
Washington, DC 20231

DISCLOSURE DOCUMENT

for

A New Type of Universal Roof Rack
for Automotive Use

Prepared by

Jeffrey J. Spira
Inventor

Prepared for

U.S. Department of Commerce
Commissioner of Patents and Trademarks
Washington, D.C.

Signed _____Date _____
        Jeffrey J. Spira

1. Introduction

The following document is offered as a Disclosure Document to the Commissioner of Patents and Trademarks. It discloses a new type of universal roof rack for automotive use. This roof rack offers an entirely new technology never before built, described, or offered. Jeffrey J. Spira is the sole inventor and developer of the rack presented here and intends to pursue the filing of a United States patent for the device.

2. Background

Many types of roof racks are currently available and offered to the public for sale. Each is designed to accommodate a specific type of cargo. There are cargo carriers, bicycle carriers, ski racks, and boat racks.

To date, no one has offered a universal type carrier that is readily adaptable to any configuration of cargo. The device presented here can be reconfigured as a cargo carrier, a ski rack, bicycle rack, or boat rack.

3. Description of the Device

The universal roof rack consists of an aluminum, rectangular subframe with conventional rain gutter clamps for attachment to the automobile. Interchangeable overframes for carrying bicycles, skis, boats, or other cargo snap onto the subframe by means of retractable pins permanently connected to the subframe. The pins are held in place with a spring arrangement to lock the overframes in place.

4. Drawing of the Rack

Following is a drawing of the universal automotive roof rack.

APPENDIX C

Secrecy Agreements

Many people believe that for agreements to be legal they must be clouded in a bunch of legal language. The rationale is that unless every sentence begins with a bunch of heretofores and whereinafters it couldn't possibly be legal. Nothing could be further from the truth. The contracts manager at my first job out of college taught me much about contracts. He insisted that the language be straightforward and simple. All of the fine print and legal mumbo jumbo are invented by lawyers to keep themselves busy—remember they get paid by the hour.

A secrecy agreement is nothing more than a written statement that whatever you discuss will remain between the two of you. No one will take any action on the discussion without further agreements. That's all it is. In fact, the more simply stated, the more general and all-encompassing the agreement becomes.

Both you and the person you're discussing your product with should sign the secrecy agreement. You should have a signed secrecy agreement **from each person present at any** meeting in which you discuss the new product or invention. It is necessary to get only one signed agreement from each person, no matter how many times you meet about the same product. If, however, you come up with another, different product, you should get a separate agreement for each product.

Here is the secrecy agreement I use. It has worked well

so far. No one has ever infringed on it, so it has never been tested in court. My lawyer has seen it and thinks it's adequate, so I don't worry about it. You are welcome to copy it word for word. You may want to have an attorney look it over to be sure. Laws concerning contracts, secrecy agreements, and other legal aspects of business vary from state to state.

SECRECY AGREEMENT

I,_____, understand that certain proprietary information about a new product development or invention will be discussed or revealed to me. I agree not to discuss this development or information with anyone not specifically included in the discussions and having signed a secrecy agreement.

I also further agree not to manufacture or offer for sale any product incorporating the proprietary information discussed without expressed written consent from (your name).

Signed: _____ Date: _____

Company: _____

Signed: _____ Date: _____

Index

125